T0092329

Springer Proceedings in Mathematics

Volume 18

For further volumes:
http://www.springer.com/series/8806

Michael Field

Claude Bruter

Editor

Mathematics and Modern Art

Proceedings of the First ESMA Conference,
held in Paris, July 19-22, 2010

 Springer

Editor
Claude Bruter
ESMA
Institut Henri Poincaré
Paris, France

ISSN 2190-5614   e-ISSN 2190-5622
ISBN 978-3-642-24496-4   e-ISBN 978-3-642-24497-1
DOI 10.1007/978-3-642-24497-1
Springer Heidelberg Dordrecht London New York

Library of Congress Control Number: 2012934047

Mathematical Subject Classification (2010): 00A66, 00B15, 35C08, 3701, 37C10, 51M10, 5301, 5401, 57M25, 5 M99

Printed on acid-free paper

Springer is part of Springer Science+Business Media (www.springer.com)

Preface

The first Conference of the European Society for Mathematics and the Arts (ESMA) was held at the Henri Poincaré Institute in Paris from 19 July to 22 July 2010, and was accompanied by an exhibition at the same Institute (see http://math-art.eu/ihp10/index.html). This volume gathers together the texts of the majority of talks held during the conference.

A large proportion of the public may still question whether one can closely link mathematics and art. In fact, that link, implicit or explicit, was established with the first creations of decorative and religious art. Great painters, whose imagination and creativity also had a rational basis, found the structural foundations of their art inside the mathematics to whose development they sometimes contributed.

The remarkable course of the symbolic sciences in the last 150 years and what it has revealed have provided us an inkling of the diversity of forms that can populate spaces, and above all ours. Because it is not bound with numbers, this diversity is infinite.

These forms, mostly unexpected and often very beautiful, cannot help but arouse the curiosity of mathematicians and artists alike. By making these forms known through their work—which allows them to reach the peoples of all countries—artists contribute in a subtle way to making everyone familiar with this wonderful and plentiful universe of new objects. As such, they make an essential contribution to breaking down the psychological barriers that still separate mathematics from some sectors of society.

In each of the chapters of this work, readers will discover works of art whose main characteristic is that they have been created by the computer brush on the canvas of mathematical rationality. The text can be interpreted as an invitation to an initial exploration of some aspects of the mathematical world, namely those that are inherent in each work: some aspects only, as there are so many different possible approaches to the universe of these abstract physics. One can for example gain access to the practice of number theory, analysis, and algebra. Even if, when reading the book, these main fields of mathematics seem at first blush to be of little benefit to geometry, in truth they have contributed greatly to enhancing the role enjoyed by geometry in its broadest sense, used to describe not only the static and inanimate but also movement and the animate.

The emphasis placed on geometry is justified by the fact that it treats both the exterior and the interior forms of objects. Some of these forms are very familiar to us, associated with physical or biological objects created by nature long ago, or even more recently by man himself. One can observe the particularly strong presence of pure mathematical forms in artifacts of the latter type, whether it be the table (square, rectangular, circular, elliptic), the cubic pedestal, parallelepiped furniture, the pyramid on a square base, or the most recent roofing technologies to shelter opera houses. The forms are in a sense the incarnation of geometry in the physical world. Conversely, we could say that geometry is an incarnation of the physical world in the symbolic world. As such, it is fitting that many articles in this work show where these two types of incarnation appear.

First, readers of this book will no doubt be mathematicians and artists. The former will most likely explore the work out of intellectual and aesthetic interest, and to better imprint their minds with the reality and knowledge of objects that they have already encountered, or even contributed to creating. But the same concern for curiosity and interest will surely motivate many artists. Insofar as the contributing authors reveal their methods and the techniques they advocate, the articles should be capable of giving stimulus to all those who would like to acquaint themselves with these modern forms of art before penetrating further.

Humanity is following and evolutionary path that should allow it, by transforming itself, to ensure its permanence before the solar fire burns our Earth. The disembodied symbolic forms that we have constructed and that we continue to create naturally escape igneous destruction. The replacement of the oil painting and the brush by the computer, of the coloured powder by the number, of the physical motif in our environment by the symbolic motif discovered and simultaneously created by the mathematician, fits suitably into this evolution. This volume provides the evidence for a scientific and artistic movement destined to assimilate such rich developments.

Paris, France   Claude Paul Bruter

Contents

A Mathematician and an Artist. The Story of a Collaboration 1
Richard S. Palais

Dimensions, a Math Movie 11
Aurélien Alvarez and Jos Leys

**Old and New Mathematical Models: Saving the Heritage
of the Institut Henri Poincaré** 17
François Apéry

An Introduction to the Construction of Some Mathematical Objects 29
Claude Paul Bruter

Computer, Mathematics and Art 47
Jean-François Colonna

**Structure of Visualization and Symmetry in Iterated Function
Systems** 53
Jean Constant

M.C. Escher's Use of the Poincaré Models of Hyperbolic Geometry 69
Douglas Dunham

Mathematics and Music Boxes 79
Vi Hart

My Mathematical Engravings 85
Patrice Jeener

Knots and Links As Form-Generating Structures 105
Dmitri Kozlov

Geometry and Art from the Cordovan Proportion 117
Antonia Redondo Buitrago and Encarnación Reyes Iglesias

Dynamic Surfaces 131
Simon Salamon

Pleasing Shapes for Topological Objects 153
John M. Sullivan

Rhombopolyclonic Polygonal Rosettes Theory 167
François Tard

Index 177

A Mathematician and an Artist. The Story of a Collaboration

Richard S. Palais

1 Appreciation

In recent years it has become *de rigueur* for an invited speaker at a conference to "Thank the organizers for inviting me." Today I can say this with more than the usual sincerity. Paris is my favorite city in the world, I have many fond memories of the Institut Henri Poincaré, and the subject of this conference is very close to my heart. So to Claude, and all the organizers who have worked hard to make this conference a success. Merci bien.

2 Introduction

Today I would like to tell you the story of perhaps the most enjoyable and stimulating collaboration of my career—my interaction with Luc Bénard. Luc is a talented creator of breathtaking mathematical art, and it really should be Luc standing here giving a talk at this conference, but unfortunately other commitments prevented him from coming to Paris at this time.

While I appreciate mathematical art, I have little talent for creating it, and what I brought to our partnership was primarily technical knowledge and experience in creating software tools for representing mathematical objects as computer-based images. True, Luc has made good use of these tools, but I feel somewhat embarrassed by the excess credit I have received for his creations; it is as if one gave partial credit for the Mona Lisa to the artisan who created da Vinci's paintbrushes.

I looked back at my early email messaging with Luc recently. It shows that in early December of 2004, Luc was using my software, 3D-XplorMath (3DXM) and

R.S. Palais (✉)
Department of Mathematics, University of California, Irvine, CA 92617, USA
e-mail: palais@uci.edu

C. Bruter (ed.), *Mathematics and Modern Art*, Springer Proceedings in Mathematics 18, DOI 10.1007/978-3-642-24497-1_1, © Springer-Verlag Berlin Heidelberg 2012

1

asked if it would be possible to save surfaces created with the program in a certain standard format (.obj), since that would make it possible for him to use other 3D graphics programs he liked a lot (Bryce and Carrara) to further process these surfaces. Since the Australian astrophysicist and computer scientist, Paul Bourke, was an expert in such matters (as well as having one of the best Mathematical Visualization websites on the Net) we asked him for help, and over the next few weeks Paul and I implemented saving 3DXM generated surfaces as.obj files with Luc as our beta-tester.

3   The Five Glasses Surface

The payoff came in the form of a spectacular New Year's Day present from Luc! (Fig. 1)

> From: Luc Benard ludev@videotron.ca
> Date: January 1, 2005 10:26:07 AM EST (CA)
> Subject: Re: surf format
> Happy New Year to you and your family

I couldn't believe my eyes! The illusion of reality was so strong that I was completely convinced by it. I immediately recognized the surfaces as having originated from my own program, 3D-XplorMath, but I nevertheless believed that this had to be a photograph of real, physical glass models!

And I wasn't the only one fooled by it. Here is a story I told to many people as I showed them Luc's image—and most of them accepted it: These surfaces were

Fig. 1 A mathematician in Murano

initially created in software, using the mathematical visualization program called 3D-XplorMath. The resulting images were given to a highly skilled artisan glass-blower from Murano, who fabricated them out of thin colored glass. The glass objects were then given to a professional photographer who placed them artfully on a glass covered walnut tabletop and took their picture.

Of course, the virtual glass blower and virtual photographer were both Luc Bénard, from Montreal, not Murano. Here is my reply message to Luc. Note that I copied to many of my friends who are interested in mathematical visualization.

To:Luc Benard <ludev@videotron.ca>
Date: January 1, 2005 4:29:23 PM EST (CA)
Cc: Paul Bourke <paul.bourke@gmail.com>, dave hoffman <david@msri.org> karcher <karcher@math.uni-bonn.de>, xah lee <xah@xahlee.org>, <matweber@indiana.edu>, Martin Guest <martin@comp.metro-u.ac.jp>, eck@hws.edu
Dear Luc,
When I looked at your attachment I could hardly believe my eyes! My first impression was that this must surely be a careful high resolution digital photograph of real glass models sitting on a real glass tabletop and made by some incredibly skilled glassblower...

One of the people I sent Luc's picture to was my daughter, Julie Palais, who is a program director for the US National Science Foundation. She liked the picture a lot and suggested the we submit it to the Science and Engineering Visualization Challenge, that each year is sponsored jointly by the NSF and Science Magazine. Luc and I decided to do that, although we felt we would be lucky to even get an honorable mention; Science Magazine is not very math oriented, tending to favour the "hard" sciences.

BUT, much to our surprise, we won First Prize in the 2006 competition, and Luc's "Five Glass Surfaces on a Table Top" became the cover illustration of the Sept. 22, 2006 issue of Science (Fig. 2).

4 Kuen's Surface

I will come back to the "Five Surfaces" later and show you a sample of all the very hard work that Luc put into producing the final image. But first let me tell you about another beautiful surface image that Luc created about a year ago. This one is of a single surface, the pseudospherical Kuen's Surface, and it is intended to be more than just a pretty picture—as I will explain, it tells the story of a highly complex historical development that stretches from Euclid's Axioms to modern Quantum Field Theory, and we decided to enter it into the 2009 Science and Engineering Visualization Competition (Fig. 3).

For 2,000 years mathematicians searched for a proof that Euclid's other postulates logically implied his Fifth or Parallel Postulate: "Through a point outside a given line there can be drawn exactly one parallel line". Finally, in 1826, Nikolai Lobachevsky showed that this was a futile goal; he constructed a geometry that satisfied the other Euclidean postulates, but not the Fifth. Indeed, in this new

Fig. 2 September 22, 2006 issue science cover

Fig. 3 Kuen's surface a meditation on Euclid, Lobachevsky, and quantum fields

geometry there were infinitely many parallel lines through the given point. Such geometries are now sometimes called "Lobachevskian", but more commonly are called "hyperbolic", and surfaces in space that exhibit this geometry are called "pseudospherical", after the simplest example, the Pseudosphere (which looks like two dunce caps held brim to brim). But just as in addition to the plane itself there are many "planar" surfaces in space, such as cones and cylinders, that obey Euclid's axioms for plane geometry, so too there are many pseudospherical surfaces besides the Pseudosphere. A famous example, is Kuen's Surface, that has been admired for its graceful beauty since its discovery 150 years ago.

One of their remarkable discoveries was that pseudospherical surfaces were in one-to-one correspondence with the solutions of a certain nonlinear partial differential equation that we now call the Sine-Gordon Equation. Usually such equations do not have explicit solutions, but Bäcklund discovered an important sequence of so-called soliton solutions that are explicit and which moreover correspond to particularly beautiful and symmetric pseudospherical surfaces. In fact, the Pseudosphere is a one-soliton surface and Kuen's Surface is a two-soliton. A century later this same Sine-Gordon Equation mysteriously resurfaced in an entirely different context: it turned out to be a model of a Relativistic Quantum Field Theory, the kind of mathematical structure on which theoretical physicists base their most advanced and sophisticated theories of the structure of matter.

Our visualization captures some of the surprising mystery of this 2,000 year saga of intimately interconnected mathematical ideas. Our eyes perhaps dwell first on the planar objects: the floor with its two sets of parallel lines, the table-top and the piece of paper with its planar sketches of the Kuen Surface seen from various directions and an esoteric formula written below them. But though that formula lies in a plane, it is in fact the two-soliton solution of the Sine-Gordon Equation, and so describes the glassy Kuen Surface lying next to it, in whose pseudospherical shape we can see reflected the planar table top and scrap of paper with its images and its formula.

Fortune smiled on us again; the judges liked our visualization and the story behind it, and they awarded us another First Prize for the 2009 competition. Luc Bénard has produced a great many mathematical visualizations of various types, and I asked him recently which ones he liked the best. Let me show you a few of his favorites—accompanied by some of his artistic philosophy in his own words. Then, in what time remains, I will try to explain a little bit about Luc's methods for creating these wonderful images.

5 Triply Periodic Surface WP

Because I love mathematic and I always find mathematical curves and surface graceful and really close to the curves and surfaces we find in the organic world I tried to use them as objects in 3D images simulating reality. Here is an example, The Triply Periodic Surface WP (Fig. 4).

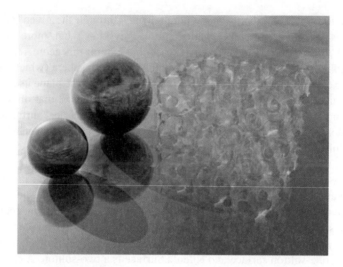

Fig. 4 Triply periodic surface WP

Fig. 5 The Wada lakes (fractal basin)

6   A Fractal Basin

Another example is the Fractal Basin: take four highly reflective tangent spheres with equal radii and centres at the vertices of a tetrahedron. You can use 3D software to replicate this reality, including some of it's physical characteristics, including the reflection of light. if you look into the space between the sphere you see a fractal. Here it is with two zooms (Figs. 5–7).

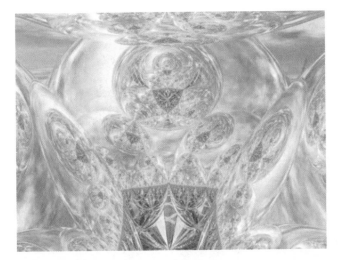

Fig. 6 Fractal basin zoom 1

Fig. 7 Fractal basin zoom 2

7 Two Cyclides

Here is another one of Luc's surface images—one of my favorites. It shows two glassy Cyclides side by side, each reflecting the other (Fig. 8).

Fig. 8 Two cylids

Fig. 9 Following Alice and the white rabbit

8 Tunnel 5B

This is a fantastic fractal! (Fig. 9)

9 Lya Poster

This is another one of my own favorites... (Fig. 10)

10 Wohlgemuth-Thayer

...and finally, the Wohlgemuth-Thayer surface (Fig. 11).

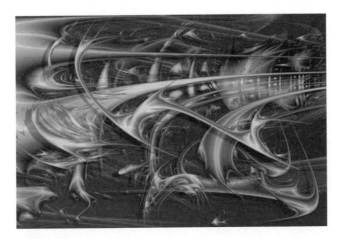

Fig. 10 Copper and gold: concertante symphony

Fig. 11 The Wohlgemuth-Thayer surface

Dimensions, a Math Movie

Aurélien Alvarez and Jos Leys

Abstract Dimensions is a 2-h animated movie, aimed at a broad audience, produced by Jos Leys, Étienne Ghys and Aurélien Alvarez. The notion of "dimension" in the mathematical sense is explained in nine chapters: "Dimension 2" talks about location on a sphere and stereographic projection. "Dimension 3" explains how 2-dimensional creatures can imagine 3-dimensional objects, which is an introduction to "Dimension 4" where we show how we, as 3-dimensional creatures, can imagine 4-dimensional objects. Next is a visual introduction to complex numbers, leading in to the Hopf fibration as an example of 4-dimensional math. As an epilogue we show a formal proof of a geometric theorem related to stereographic projection. Through this film, the authors wanted to show that math does not need to be "dry", but that math can produce beautiful imagery. In order to reach as wide an audience as possible, the film is a non-profit project. The DVD has a low price, and the films can be downloaded free of charge from an internet site featuring additional information on the subjects of the film. Furthermore, the film has a "Creative Commons" license, which allows copying of the film (provided there is no commercial gain). The film is available in 8 commentary languages and 20 languages for the subtitles.

1 Introduction: The Start of a Film Project

Early 2006, Jos Leys and Etienne Ghys, research director et the Ecole Normale Supérieure at Lyon, made their acquaintance on the internet: Ghys was looking for some images to illustrate a general-public talk he was giving, and stumbled upon

A. Alvarez (✉)
Maître de conférences, Université d'Orléans, 45067 Orléans cedex 2, France
e-mail: aurelien.alvarez@univ-orleans.fr;

J. Leys
Ingénieur, Visualisation mathématique, Antwerp, Belgium
e-mail: jos.leys@pandora.be; http://www.josleys.com

C. Bruter (ed.), *Mathematics and Modern Art*, Springer Proceedings in Mathematics 18,   11
DOI 10.1007/978-3-642-24497-1_2, © Springer-Verlag Berlin Heidelberg 2012

Fig. 1 DVD cover

Leys'website [1] where there are a large number of mathematics based images and animations. Later that year Ghys was to give a plenary talk at the 2006 ICM in Madrid, and he proposed to collaborate on the making of a series of images and animations to illustrate this talk. An intensive exchange of emails followed, with Ghys' instructions and Leys' test images, culminating in a successful plenary talk on "Knots and Dynamics" [2,3].

During this project, on numerous occasions, Ghys had to explain advanced math concepts to Leys who is a non-mathematician, and from this came the idea for a film (Fig. 1): explaining a set of math concepts in simple terms to as wide an audience as possible.

2   The Scenario

We chose to explain the notion of "dimensions" in the mathematical sense. The choice followed naturally from the work on "Knots and Dynamics" where a lot of the math was 4-dimensional, and furthermore we thought that higher dimensions would have a lot of appeal. The film consists of nine chapters of 13 min each, which is a good length of time for eventual classroom use.

"Dimension 2" talks about location on a sphere and stereographic projection. (We are making use of the latter in all the subsequent chapters.) "Dimension 3" explains how 2-dimensional creatures can imagine 3-dimensional objects, which is an introduction to "Dimension 4" where we show how we, as 3-dimensional creatures, can imagine 4-dimensional objects. As 4-dimensional objects we show

Fig. 2 4-dimensional
polytope

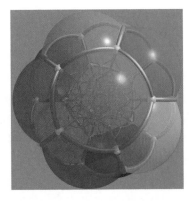

a series of 4-dimensional polytopes such as the tesseract or the 120-cell (Fig. 2). Next is a visual introduction to complex numbers, leading in to the Hopf-fibration as an example of both complex numbers and 4-dimensional math. As an epilogue we show a formal proof of a geometric theorem related to stereographic projection.

There are some still images from the film in the paragraphs below, in no particular order.

3   The Production

All of the images and animations were done in Povray [4], a free raytracing program. At 25 frames per second, we needed to render a total of about 175,000 images sized at 800 × 600 pixels. The present version of Povray does not allow multiple core processing, so the rendering of the images is relatively slow. We managed to put to work a whole series of computers at the "Pôle Scientifique de Modélisation Numérique", the computing centre at the École Normale Supérieure in Lyon where both Ghys and Alvarez resided. Each computer was given a series of images to compute, which were then collated by Alvarez to be turned into Quicktime animations using Final Cut Pro, a video editing program.

The music was chosen from royalty free sources on the internet, and from compositions by Florent Ghys, a nephew of Étienne Ghys.

4   A Non-profit Project

In order to reach as wide an audience as possible, we quickly concluded that this had to be a non-profit project, so the film could be accessible to all. This means selling the DVD at a low cost which mainly covers expenses, willingness to give out DVD's for free to appropriate organisations (as an example we provided free copies

Fig. 3 Dupin cyclide

for the participants at a Math Olympiad, and to a convention of French math teachers) and a simple licensing scheme. We chose a Creative Commons [5] license which allows copying without commercial gain with proper attribution.

The ENS at Lyon graciously provided a small grant to cover our startup costs, mainly for the production of the DVD's. (There was little or no interest from other institutions or from publishers.)

In order to keep costs as low as possible, we used mainly free or open source software for the entire project, and depended heavily on unpaid volunteers for commentary voices in different languages, and for making translations. In fact we attribute the success of the film project for a large part to the availability in multiple languages. The film is available with commentary voice in 8 languages and subtitles are available in 20 languages.

5 The Website [6]

We felt from the start that we needed a website for this project. Firstly because it is just about the only way we had to let the world know that the film exists at all. For this aim we could just have made a one-page site with a small description of the film and a button to buy the DVD. However we felt that we could do a lot more. There is now on the website a page for every chapter of the film with additional information on the different topics, pages where the film can be downloaded freely (several mirror sites across the world are available) in various formats (Quicktime high resolution, Ipod format), directly or by Torrent, pages where the film can be watched online, and of course a page where the DVD can be ordered. (The price of 10€ covers the cost of the DVD as well as worldwide shipping.) Through the hard work of dozens

Fig. 4 Mandelbrot fractal

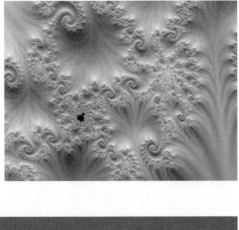

Fig. 5 Stereographic
projection

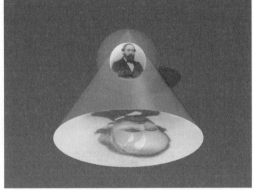

of volunteers, the website is now available in 10 languages (French, English, Dutch, German, Arabic, Russian, Chinese, Japanese, Spanish and Portuguese).

6 Results

Since the day of the opening mid-2008, the website has been visited close to a million times from just about every country in the world. The number of downloads of the film is difficult to estimate because of the propagation via Torrents, but we see that the site at Lyon has a bandwidth of a few hundred gigabytes per day. To date we have had 12,000 DVD's produced of which about two thirds have been sold, and one third given away.

In June 2010, the film received the "Prix d'Alembert", a bi-annual price given by the French Mathematical Society to what they judge is the best mathematics divulgation project of that period. More importantly, we still receive enthusiastic reactions from all over the world on a daily basis.

7   Conclusion

We believe that with his project, we have proven that it is possible to produce a good math divulgation film with very limited resources, provided there is a team of enthusiasts willing to spend a considerable amount of time. Success is only guaranteed with a good team. In our case it consisted of two mathematicians and an engineer, and we found that this collaboration works very well.

In fact, this same team has plans for a new project.

References

1. Mathematical Imagery: *http://www.josleys.com*
2. Étienne Ghys on *Knots and Dynamics* at the 2006 ICM in Madrid: http://www.mathunion.org/Videos/ICM2006/muster.php?ghys2006
3. Étienne Ghys and Jos Leys, *Lorenz and modular flows: a visual introduction.* Feature column of the AMS: http://www.ams.org/samplings/feature-column/fcarc-lorenz
4. See http://www.povray.org
5. Creative Commons: Attribution – Non commercial – No derivatives 3.0 http://creativecommons.org/licenses/by-nc-nd/3.0/
6. *http://www.dimensions-math.org*

Old and New Mathematical Models: Saving the Heritage of the Institut Henri Poincaré

François Apéry

1 Introduction

During the inaugural conference of the very new european society dedicated to the interaction between mathematics and arts (ESMA), which was held at Intitut Henri Poincaré (IHP) in Paris in July 2010, I was given the opportunity to tell about the state of the substantial collection of mathematical objects, which are displayed in the IHP library or are stored in the underground reserve, because of poor condition or which are in process of being catalogued.

This collection has over the century, been subject to various events, including lack of interest, left on shelves or some being moved to Palais de la Découverte (and incidentally photographed by Man Ray in the 1930s). However, over the past decade, there has been a renewed interest thanks to those people, who have been encouraged by a more favourable climate after the wave of formalism that had swept through all components of scientific activity, especially in France. After this neglect, we are seeing some beautiful objects, that, abused by some damage of the years, one thinks these are revealing knowledge about the prehistory of geometry.

Although these models have permeated through geometry lessons at all times, students have not always been able to recognize them from their formal description nor from rough sketches on the blackboard.

Thanks to technical progress of computer science, virtual images, in other words, potentially concrete objects, have been able to be portrayed in a realistic way giving rise to a regained attention to two and three dimensional pictures, and, as a result, to a need to have the objects at hand. For the understanding of a mathematical object is greatly facilitated by a back and forth between the necessary abstraction effort and the handling of the physical object, the transition through the computer screen being a go-between no more and no less essential than a stair.

F. Apéry (✉)
F.S.T, 68093 Mulhouse Cedex, France
e-mail: francois.apery@uha.fr

C. Bruter (ed.), *Mathematics and Modern Art*, Springer Proceedings in Mathematics 18, DOI 10.1007/978-3-642-24497-1_3, © Springer-Verlag Berlin Heidelberg 2012

However, there is a problem. The models in Palais de la Découverte and the IHP cannot be handled by visitors at the risk of damage, so they are either locked in cupboards inside the junk room or treated as works of art and placed in showcases. However, unlike works of art, they are intended to be reproduced for their main function is didactic. The main reason for these objects is to give the thought of ordering or making copies, for either personal or student use.

The famous Martin Schilling collection has been sold to many universities at the beginning of the twentieth century, and now one can see almost the entire collection at IHP. They have come from the geometry chair of the former Sorbonne and Gaston Darboux used it in his lectures. Most of the wooden models illustrated the descriptive geometry exercises by Joseph Caron at École Normale Supérieure (see for instance [2]). Some models were given in the late forties to IHP by El-Milick, a mathematics professor, who was also a gifted amateur sculptor.

However, this could be said, all this is nothing but past feeling, past mathematics, old-fashioned outdated things. Nevertheless, if the language and the methods evolve, checking a statement on an explicit example is still full of interest, and further, what could be more delightful as to deal with a touchable object able to elicit the geometric sense. In addition, and this is the main point, a collection should not be static but enriched by new models suggested by working mathematicians. Proposing two wire models (a Boy surface of degree six and a Morin surface of degree eight (Fig. 1)) presently shown in the IHP library, I wished to lead by example.

The virtual issue remains. Today, the computer is present everywhere, from conception to final realization. So, why not stop at the virtual object, since it can be thoroughly explored on a monitor, thanks to more and more efficient 3D software? Wildlife documentaries are no substitute for a zoo. It is possible to explore wildlife superficially on the screen, but animals live, become ill, procreate and the zoologist is more concerned with the real animal than its digital glint.

In spite of what a shallow view might let think, a mathematical object, pure abstraction, often stems from real intuition and, as a feedback, can lead to a physical realization that, in addition, especially when it is the unique solution of a natural problem, enjoys plastic qualities that artist souls, like surrealists in the thirties, can take over. One can not refrain from combining visual and tactile pleasure, without

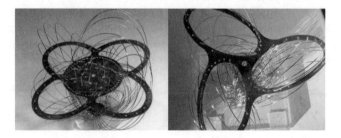

Fig. 1 Two wire models realized by the author

depriving oneself of the feeling of substance and materials. Virtual imaging is excellent but it is not all. It is sometimes difficult to develop a correct mental image without having the object to hand.

The full IHP collection consists of about 500 models that one can roughly classify according to the material and the richness of the class: glass, plastic material, cardboard, wire, wood, sewing thread on a rigid structure, plaster. The Conservatoire National des Arts et Métiers has created its own Museum, why the IHP wouldn't valorize its patrimony? It is now in process. Here, below, there is an example in each class as well as an example of what present technical devices enable one to realize. Pictures of Figs. 2, 3, 5–11 are the work Sabine Starita.

2   The Klein Bottle

The famous Klein's surface (Fig. 2) doesn't exists in our space, that is, in mathematical words, cannot be embedded in the space $\mathbb{R}^3$.

Therefore, we have to do with an immersed image, i.e. with possible self-intersections. As a result, there are several possible models, the actual value is two if we confine our attention to immersed surfaces, in other words, to immersions up to source and target diffeomorphism and a regular homotopy of the target. The one we are speaking about, in glass, justifies well its name of Klein bottle, although according to some historians, its origin would rest on a confusion between the german words "Flasche" and "Fläche".

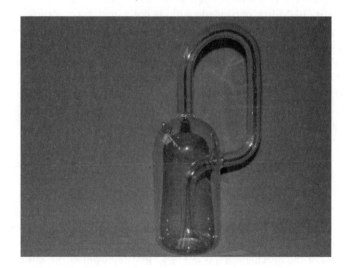

Fig. 2 Klein bottle

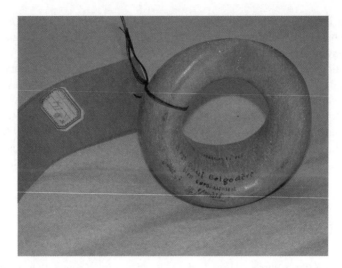

Fig. 3 The one-sided cyclid

3   The One-Sided Cyclid

This model (Fig. 3) has been realized in July 1947 by Maurice El-Milick and subsequently offered to Paul Belgodère in order to enrich the collection placed in his charge as the librarian of the IHP.

El-Milick produced a number of geometric models, some of them becoming visible on the picture of Fig. 4. We see noticeably his one-sided cyclid as well as a crosscap in plaster which now belongs to the IHP's collection.

Incidentally, El-Milick one-sided cyclid is not a cyclid, i.e. a surface of degree four doubly passing through the umbilical, but a surface of degree six the shape of which reminding that of the ring Dupin cyclid (Fig. 5).

It doesn't seem that El-Milick identified his one-sided cyclid as a stable image of the Klein bottle, that is the image of the Klein bottle by a mapping which is not an immersion. Indeed the only singularities are two pinch points, also called Withney umbrellas, connected by a self-intersection line segment. Up to ambient isotopy, the mapping remains unchanged by a small perturbation.

4   The Poinsot Great Dodecahedron

Here, we are speaking of a cardboard model coated by a polish (Fig. 6). It represents the great dodecahedron discovered by Louis Poinsot in 1809.

The edges of the polyhedron are in black. Red lines figure the self-intersection lines of the faces. It has 12 faces, 30 edges and 12 vertices, so that its Euler-Poincaré

Fig. 4 Maurice El-Milick

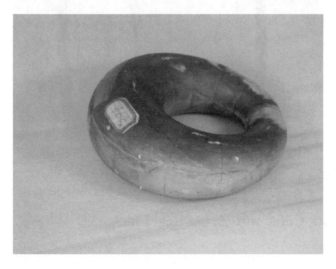

Fig. 5 Ring Dupin cyclid

characteristic is equal to -6, and consequently, since it is orientable, it represents a cell decomposition of the closed orientable surface of genus 4.

Fig. 6 The great dodecahedron

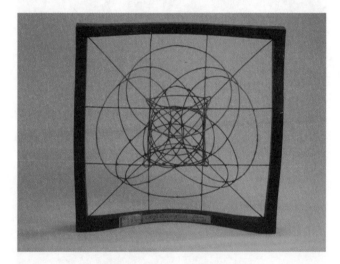

Fig. 7 Smooth cubic with seven real lines

5 Smooth Cubic with Seven Real Lines

On this wire model of surface realized by Joseph Caron on June 10, 1912, we only see a few noticeable curves traced on it (Fig. 7).

The surface, the equation of which being

$$Z(X^2 + Y^2 + Z^2) + 2(X^2 - Y^2) - 16Z = 0,$$

is a smooth cubic, that is without any singular point. In 1858 Ludwig Schläfli got the idea to classify such surfaces according to their number of real lines: provided that the cubic surface is not ruled, then it falls into only four different types in reference to the value 3,7,15 or 27 of that number. In the present case, the surface contains seven real lines, six of which being visible, and the seventh being cast aside at infinity. In addition, the surface is generated by ellipses, some of them are circles and are shown on the model. The choice of the curves materialized by the wire preserves the invariance under the group of space isometries leaving the surface globally unchanged, which is isomorphic to the 8-order dihedral group and generated by the two following linear transformations:

$$\begin{pmatrix} 1 & 0 & 0 \\ 0 & -1 & 0 \\ 0 & 0 & 1 \end{pmatrix}, \begin{pmatrix} 0 & -1 & 0 \\ 1 & 0 & 0 \\ 0 & 0 & -1 \end{pmatrix}.$$

Although it has a very similar shape (actually it is isotopic through an ambiant isotopy of $\mathbb{R}^3$), this cubic surface must not be mixed up with the ring parabolic cyclid (Fig. 8, Serie X no. 5 Schilling collection) of equation

$$Z(X^2 + Y^2 + Z^2) + 4(X^2 - Y^2) - 16Z = 0.$$

As a matter of fact, the latter has two double lines. Both of them separates the space into two isometric subsets which are swapped to each other by the half-turn rotation about the axis X, Y and $Z = 0$.

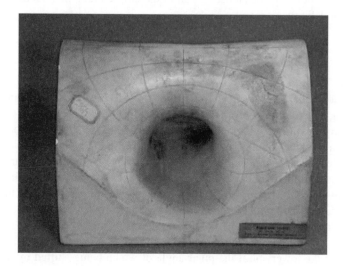

Fig. 8 Ring parabolic cyclid

Fig. 9 Dandelin model

6 Dandelin Model

The educational aim is clear (Fig. 9).

On a wooden cone divided into two parts in order to make two inscribed wooden spheres visible, one notices, on one hand, a brace rod figuring a generating line of the cone, and, on the other hand, a metallic ellipse whose plane is tangent to both wooden spheres. The points of tangency on the spheres materialize the focus of the ellipse (first part of Dandelin theorem), whereas the touching circles of the spheres with the cone define planes meeting that of the ellipse along its directrices (second part of Dandelin theorem).

7 A Model of Descriptive Geometry

This model built by Joseph Caron in July 1914, was obviously intended for illustrating a lecture on descriptive geometry (Fig. 10).

The purpose of this field, today missing, and created by Gaspard Monge was to represent a 3-dimensional object by two orthogonal projections on two orthogonal planes. The object we are dealing with is the intersection of two ruled surfaces. The first one is a right cylinder the base of which is a logarithmic spiral and its generating lines are materialized by a green sewing thread. We get the second one by dilating the spirals and corkscrewing the cylinder, in the same way one produces a one-sheeted hyperboloid from a revolution cylinder, that is by connecting (with a red sewing thread) the generic point of the upper spiral, with the point of the lower spiral the polar angle of which being shifted by a constant. The intersection curve of the two surfaces occurs as a sequence of small pearls.

Fig. 10 Intersection of two surfaces

Fig. 11 Sievert surface

8 Sievert Surface

Here, we are speaking about a plaster model of the Schilling collection designed by Georg Heinrich Sievert in 1886 in view to produce an explicit example of Joachimstal surface parametrized by elementary functions and which is not of revolution. One distinguishes three bulbs and one guesses a fourth one hidden behind. They are engraved with a network of lines of curvature (Fig. 11).

Some are plane, and the others are spherical. That is the reason of the presence of smooth round spheres which figure four spheres of geodesic curvature (two big and two small) meeting the surface orthogonally along curvature lines, and, so doing, enforcing a theorem due to d'Ossian Bonnet (see [3]).

9 Two Recent Models

The first one (Fig. 12) is a stable model of the Klein bottle with 12 pinch-points (see the section on the one-sided cyclid).

We get it as connected sum of two roman Steiner surfaces. It has been made by Stewart Dickson in 2005 on a laser prototyper. It has been named Etruscean Venus by George Francis [4].

In the same line of thought, the next picture (Fig. 13) shows a Boy surface of degree six generated by a family of ellipses and opened by windows in order to make visible the whole structure [1].

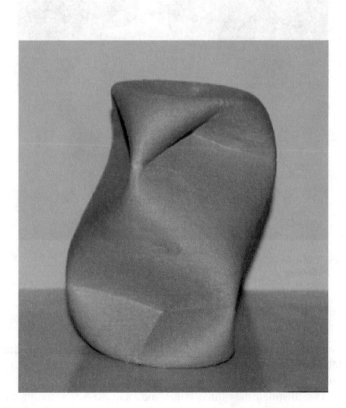

Fig. 12 Etruscean Venus

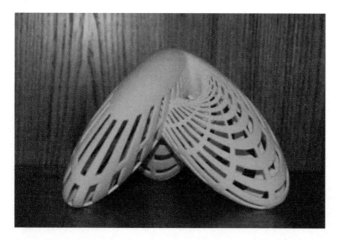

Fig. 13 Boy surface of degree six

It is the same surface as on the right in Fig. 1. It has been made by Gregorio Franzoni in 2008 on another kind of laser prototyper (see the website www. mathshells.com).

Both models, not yet belonging to the IHP collection, show what on can imagine to realize in order to update it, thanks to the present-day technics.

References

1. Apéry, F.: Immersionen der reellen projektiven Ebene in $\mathbb{R}^3$. Math. Semesterberichte **591**, Springer-Verlag (2011)
2. Brette, J.: La collection de modèles mathématiques de la bibliothèque de l'IHP. Gazette des mathématiciens **85**, 4–8 (2000)
3. Fischer, G. (ed.): Mathematische Modelle. Vieweg, Braunschweig (1986)
4. Francis, G.K.: A Topological Picturebook. Springer, New York (1987)

An Introduction to the Construction of Some Mathematical Objects

Claude Paul Bruter

Abstract In order to understand and to reconstruct the shape of many objects of the geometric world, mathematicians have focused their attention on singularities and deformations. The purpose of this article is to present these usual topological concepts and tools to artists being a priori unfamiliar with mathematics, with the hope that new beautiful creations will appear in the artistic world.

1 Generalities

1.1 Introduction

Artists have often used various kinds of objects in their composition. Mostly images of real objects. All had in common some level of style and representation, pushing forth in their mind the process of stylization. Some created shapes of an abstract nature, as in Egyptian friezes, Roman tilings, or Celtic knots. In essence, there have been two recurring threads in the art of decoration, spirals and tilings [1].

Artists did not develop mathematical theories from their constructs except during the Aeschyleus time and the Renaissance. However, we may consider them as precursors of those theories. Nowadays, many artists are familiar with the larger classes of mathematical objects that appear in their compositions. Using the power of their imagination and the tools with which they defined those objects, could contribute to enrich the catalogue of mathematical objects and the content of their work for our contemporaries and future audiences.

The purpose of this presentation is to give a general introduction to various construction principles frequently used by mathematicians and artists as well. It will

C.P. Bruter (✉)
ESMA, C/o Institut Henri Poincaré, 11, rue Pierre et Mare Curie75231, Paris Cedex 05, France
e-mail: bruter@me.com; bruter503@gmail.com

C. Bruter (ed.), *Mathematics and Modern Art*, Springer Proceedings in Mathematics 18,   29
DOI 10.1007/978-3-642-24497-1_4, © Springer-Verlag Berlin Heidelberg 2012

only focus on the first level of those principles to avoid technical difficulties and complex mathematical theories.

However, we hope some artists will find the following information useful for their work. We will address in particular those who, for whatever reason, are not using the full power of the computing environment and professional software. Salvador Dali's pictorial legacy is a very good example of the way mathematical knowledge can be used by an artist to create paintings of indisputable originality. Artworks related to the cubism expression, such as with the imaginative Chagall, are another example of the potential of human creativity.

Our approach will focus on the topological perspective only, as this rather qualitative choice carries an intrinsic insufficient element of precision. Algebraic and analytical approaches avoid this difficulty but need some mathematical knowledge and training. An advanced introduction to some of those useful techniques was published recently in the Mars issue of the Bulletin of the American Mathematical Society [2]. It should be noted that these structural and quantitative techniques have yet to be developed to reach the full potential of the shapes that are suggested by the qualitative approach. Mathematicians could enrich this qualitative approach with quantitative, numerical controls of all the deformations implied in the construction of objects. As to the qualitative approach, the specific mathematical objects studied by George Francis in *A Topological Picturebook* [3] may also be useful to advanced readers.

We presuppose that all readers understand the notion of (topological) dimension[1] of a space. Since this expose addresses artists and not mathematicians, most other terms will not have the same precise semantic of a larger, more global (mathematical) definition. That choice should allow for a more immediate understanding of the narrative.

In a broad sense, potters, sculptors and, in a more subtle way, painters will be encouraged to find that they share with mathematicians involved in geometry the use of similar processes, deformation and attachment being the most common.

As for deformations, we shall divide them into two opposite types: expansion and restriction, each one being in turn divided into two opposite types, singular or regular.

1.2   Shapes as a Restricted Class of Mathematical Objects

In general terms, mathematical objects can be classified as follows: objects that can be visualized and objects that are too abstract and too general to be a priori physically represented, such as categories or functors.

---

[1] The topological dimension of a point is 0, of a line, 1, of a surface 2, of our usual space 3, of the space-time 4, etc.

In the following, we will focus only on the class of mathematical objects of the first kind.

Definition: Such mathematical objects will be given the shorter and generic name of *shape*.

Mathematical artists use to deal with immersed or embedded 1, 2 or 3-dimensional shapes. Shapes of dimension one are mostly polygons such as triangles, classical curves as parabolas, knots or fractal lines. In dimension 2, the shapes derive mostly from polytopes, tessellated surfaces, minimal surfaces, topological surfaces, algebraic surfaces. They are often used in their strict mathematical definition and representation as in convenient deformations.

Using a general definition of mathematical objects, artists create their own shapes. To draw them, they use first a pencil or a pen or more directly, numerical symbols.

There are two kinds of numbers: static ones, and dynamic ones.

– A static n-dimensional number is a collection $(x_1, \ldots, x_n)$ of n usual real numbers. Indeed, from the dynamic point of view, it represents a translation.
– A simple dynamic n-dimensional number is slightly more general. It represents a dilatation coupled with a rotation in an n-Euclidean space. When $n = 2$, it is simply a Chuquet number also called a complex or a mixed number. One can do standard algebraic geometry with simple dynamic n-dimensional numbers that are adapted to the control of deformations, such as conformal and quasi conformal deformations in particular.

We will address first those who opt to use the pencil and brushes.

1.3 Characteristic Features of a Shape

When one observes an object, the eye follows a trajectory, a path going from some significant part to another significant part of that object. That path has been defined as the "skeleton of the percept". The parts will be named singular.

In other words, one of the most important characteristic of a shape is its set of singularities: intrinsic singularities and singularities seen from the point of view of the observer. Thus, any change on the singularities of the object is of great significance.

It should be noted that in general, the appearance or disappearance of a singularity or its modification has an important impact on the representation of the object since local or even global curvatures can be drastically changed.

Curvatures are also fundamental characteristic features of a shape. Dramatic changes in curvatures happen on singular parts.

These singular parts play an important role in the setting up of a work of art. They have both a strong significance and a semantic significance. The thorn of a rose, the fang of a vampire, the point of a sword, the blade of a knife are typical singular shapes frequently inspiring fear or violence. They represent symbolic tools of protection and attack, of preservation of one's integrity living in a dangerous

Fig. 1 Canadian Eskimo Art

world. Because they are assigned an essential role in the preservation of the self, they hold in our mind an ambiguous status. Acute angles are somehow aggressive. Drawings and paintings with many straight lines and sharp angles carry a connotation of rigidity and coldness. They also have something in common with the outline of a skeleton and define a kind of structural representation of the object they are associated with.

Definition: A *singular part* within an object of standard dimension n is a subobject of dimension k strictly less than n, which may be connected or not.

Such a singular part is characterized by some local properties of extremality of the object, like for instance the top of the head or the tip of a nose (Fig. 1).

According to our definition, a singular part can only be a point in a 1-dimensional object curve, while in a surface, which is a 2-dimensionnal object, a singular part can consist of points and/or of portion of curve.

The neighbourhood of the singular part is of course said to be regular. What makes the difference between a regular domain and a singular one?

Let us move our finger on the sculpture, first on the head from the left to the right: we observe that the finger goes up until it reaches the top of the head, then it goes down. Thus, there is a dramatic change in the move of the finger when it reaches the top of head creating a singularity: it was going up, now it goes down. If we draw a tangent to the trajectory followed by the finger, the slope of this line is "up" (positive) when the finger is on the left, is "down" (negative) when the finger is on the right (Fig. 2).

Fig. 2 Going up and down

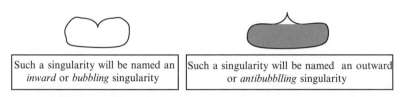

Such a singularity will be named an *inward* or *bubbling* singularity	Such a singularity will be named an outward or *antibubblling* singularity

Fig. 3 Bubbling and antibubbling singularities

It is this kind of general phenomenon which is used to characterize a singular part: "dramatic", "catastrophic", "sudden" changes in the directions of the tangent lines or planes in the immediate surroundings of that singular part.

In a singular point, a discontinuity occurs into the sign or into the value of the slopes of tangents belonging to the outer edge of that point.

As a result, most of these singular parts can be obtained by a pinching process of the set of internal modifications and deformations of the object as described below.

2 Internal Modifications

2.1 Pinching

Definition: We shall call *pinching*, denoted by $P_{n \to k}$, the process of smooth deformation that transforms a regular domain of the shape of dimension n into a domain of a singular part of dimension k less than n.

Indeed, a pinching process on any part of the object is obtained by creating discontinuities into the sign or into the value of the slopes of tangents belonging to the surroundings of that part.

On a curve, pinching occurs at points, while on a surface pinching can occur at points or along curves taking the character of singular parts.

Example 1: Geometrical (Figs. 3 and 4)

Example 2 (Fig. 5): Using antibubbling singularities, and indeed some other topological tools, could an artist have created the following lilac flower? (Fig. 6):

Example 3: An other geometrical object, the Eigthy.

Take a tube, an hollow cylinder. You can pinch it along a generatrix, in a bubbling way, or in a antibubbling way. The original generatrix becomes a singular line. Note the important reversible character of the operation (Fig. 7).

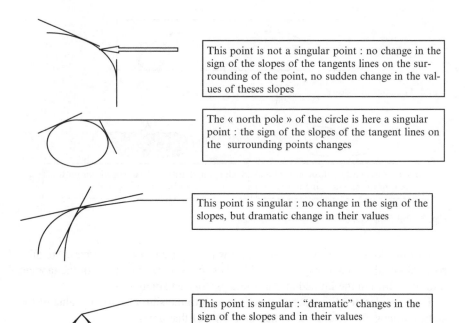

Fig. 4 Regular and singular points

Consider the bubbling cylinder. You can make a second antibubbling pinching in such a way that the two singular generatrix can be confounded. You can even pinch locally the result into a singular point and paint it to obtain the following figure called the Eigthy (Fig. 8):

Other topological techniques allow to construct this figure.

Before leaving the usual notion of singularity, let us observe another kind of phenomenon. Suppose that some part of a curve begins to vibrate more and more strongly. It may break into very small pieces, getting smaller and smaller until it finally dissolves into points, making a continuous, quasi continuous or even a discrete set. We would call this phenomenon a *fractal singularisation*, and its result a *fractal singularity* (Figs. 9 and 10).

2.2 Inflations

There are two kinds of inflations, singular inflations and regular inflations.

The singular inflations are the most interesting: they are attached to metamorphosis and the creation of new unexpected shapes and behaviours.

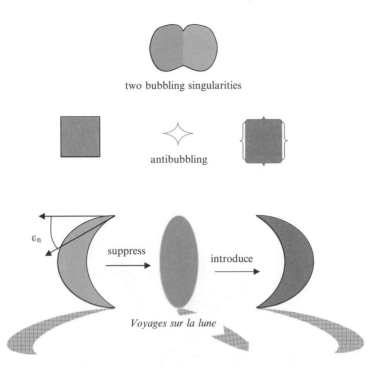

Fig. 5 Examples of births and evolutions of singularities

Fig. 6 Orchid: *Paphiopedium venustum*. On the *left* and *right* of the flower, two antibubbling singularities

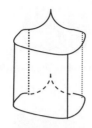 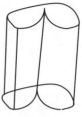

Standard cylinder   antibubbling cylinder   bubbling cylinder

Fig. 7 Singular lines on a surface

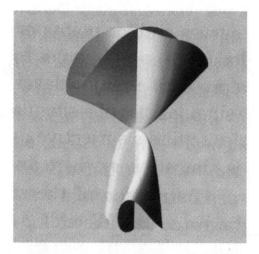

Fig. 8 The Faber-Hauser Eigthy

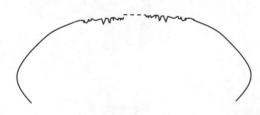

Fig. 9 Fractal singularities

Regular inflation are used in art to trigger emotion on the mind of the observer and emphasize didactic messages: El Greco (in most of his works) stretches his characters' features to express the yearning of the soul for God. Honoré Daumiers' caricatures and Hieronymus Bosch's figures, also represent two kinds of inflation.

Fig. 10 Mountain
singularities

2.2.1   Singular Inflations

The pinching process has a converse we will call singular inflation. The use of the
words "singular inflation" is restricted here to singularities. We are going to inflate
singularities.

Definition: Denoted by $I_{k \to p}$, a *singular inflation* transforms a k-dimensional
part into a p-dimensional part (k < p), the inflated part.

Done suddenly, it will be here called a *blowing-up*.[2]

The inflated part is related to its singular generative part by a few properties
among which a trivial but fundamental property: the singular generative part
can be obtained from the inflated part by a continuous deformation leaving invariant the topological properties of the successive deformed parts, properties having
an exceptional character for the singular part.

Note: A pinching process compresses a part of dimension n into a part of lower
dimension k. Usually, there might be many acceptable such parts of lower dimension, even when we restrict k to n−1. Additional constraints can of course reduce
the amount of possibilities.

A similar assertion can be done concerning singular inflations: for instance
a point can be inflated into a segment, a circle, a sphere, or a 2-disk, a 3-ball, etc.
The general procedure to inflate an object is of course to go step by step and
increase the dimension step by step (Fig. 11).

Example 4: In these two examples symbolized by $I_{0 \to 1}$, a singular point blows
up into a circle or in a shape having the same topological properties as the circle
(Figs. 12 and 13).

Example 5: Note that in the two preceding cases, the singular point could even
blow up into a disk, a two-dimensional object, which in turn can be flat, bubbling,
antibubbling, and mixed (Fig. 14).

---

[2] "Blowing up" is a standard mathematical term (in French "éclatement"). Mathematicians call the
inflated part the "blowup" of the singular part.

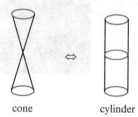

Fig. 11 Singular inflations into circles of various sizes of one point

cone   cylinder

Fig. 12 A singular point inflates into a circle

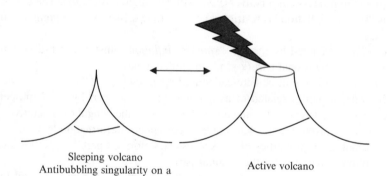

Sleeping volcano
Antibubbling singularity on a   Active volcano

Fig. 13 A singular point inflates into a circle

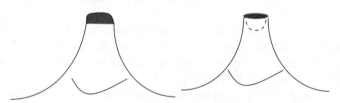

Singular inflation into an antibubbling disk   Singular inflation into a bubbling disk
giving rise to a dome   giving rise to a crater

Fig. 14 Some singular inflations into a two-disk of the singular point or of the circle of the Example 4

Note that when the 2-disk is not flat, it can be deformed into a tube open on one side, whose axis can be any non closed curve. The classification of knots can be used to classify these curves.

Fig. 15 The dome of Fig. 14, left, can be deformed into these topologically equivalent tubes

Fig. 16 Singular inflations of a singular point with multiplicity 2, left two bubbling, right one bubbling and one antibubbling

The artist will enjoy the drawing of such a tube nicely winding from and around the surface and the circle on which it arises. From that tube, fanciful horns are sometimes growing up or vanishing, throwing fantastic beams of light which illuminate an unexpected choreography.

The tube may simply show an undulation, that disappears when the viewer moves away from the initial singularity (Fig. 15):

Example 6: In the previous example the singular point blows up into a single 2-disk. But we may also consider the possibility that this point blows up into *a multiplicity* of disks having the same circle as boundary. There are particular cases of such an occurrence. We will select the simplest one, when the multiplicity is two, one disk being bubbling, the other one antibubbling: we get a 2-sphere S^2 since the 2-sphere can be constructed by identifying the circles which borders two disks.

Indeed, the singular point can be blown up into the 2-sphere since, conversely, that sphere can be continuously shrink into a point. Note that again, from our topological point of view, the sphere can be replaced by any shape having the same topological properties (Fig. 16).

If exceptionally there is a continuous infinity of this type of double expansion, the spheres may completely fill a bowl represented here by the symbol D^3.

Example 7: Let's take an orange as the physical representation of such a bowl and the 2-sphere S^2 as its boundary. We may consider that boundary as the singular part of the bowl.

There are two main ways to inflate the 2-sphere, either inwards towards the centre of the bowl, or outwards. Any such singular inflation may be partial or complete. A complete inwards inflation of the 2-sphere is the 3-ball D^3. A complete outwards

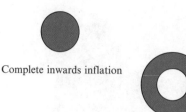

Complete inwards inflation

Complete outawards inflation

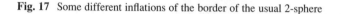

Section of a partial inflation through the centre of the sphere

Fig. 17 Some different inflations of the border of the usual 2-sphere

inflation fills up the usual 3-space leaving an hole that the previous 3-ball would fill up. A partial inflation creates an object that looks like the 3-ball with an hollow in its interior (Fig. 17).

This partial inflation is also called a thickening. We would rather call it a *standard thickening*. It is usually described as a Cartesian product: let B the boundary, I an interval, the standard thickening is described as the product B × I.

For instance, if T is a hollow cylinder or a tube without thickness, T × I will denote a tube whose local thickness is I; or if D^3 is a 3-usual ball whose border is an usual sphere of radius 1, the standard thickness of D^3 will be a 3-ball whose radius is 1+ the length of the interval I.

More generally, let C be any other object: the Cartesian product B × C may be understood as a thickening of B through C.

2.2.2   Regular Inflations and Desinflations

These transformations can be global or local. Optical illusions, anamorphoses introduce increasing or decreasing local sizes through lengths and twists. The perspective theory has formalized some of those transformations of size that do not change the topological properties of the shape. Inflation is an important component of visual communication expressing power, will, and hopes.

Foldings are frequently used as a first step in the process of developing transformations.

2.3   Folding

Definition: *Folding* is the operation by which one can act on a part of an object and change the local curvature along the points of that part, and eventually the size of that part.

There are two types of foldings: continuous foldings, and singular foldings as in the art of origami—a paper folding methods. The folding of a domain is continuous when the direction of the perpendicular to the tangent line or plane along any

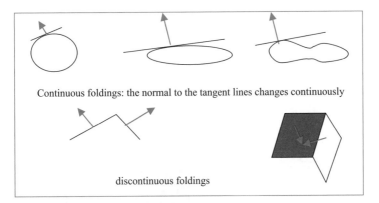

Fig. 18 Simple geometrical illustrations of the two kinds of foldings

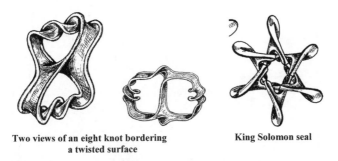

Two views of an eight knot bordering a twisted surface

King Solomon seal

Fig. 19 Three knots drawn by George Francis

transversal line to the domain changes continuously. If a discontinuity appears somewhere in that change of direction, the folding is locally singular (Fig. 18).

Example 8: In 2-dimension, when drawing on a sheet of paper, the folding of a line can use not only changes of its length but its rotations as well. In 3-dimension, the folding of a line uses changes of its length and *twists* which are couples of simultaneous rotations in two non parallel planes. Note the semantic and artistic importance of twists, expressing at once force and motion as the works of Michelangelo and El Greco titled *Laocoön*.

Examples of geometrical twists as in the work by George Francis [3] (Fig. 19):

2.4 Cutting and Opening

Definition: *Cutting* is an act of separation, of disconnection, done along any k-dimensional part of an object of dimension n (>k).

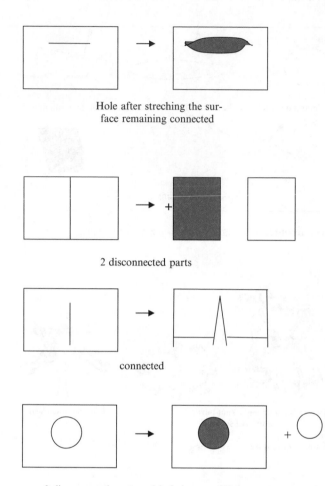

Hole after streching the surface remaining connected

2 disconnected parts

connected

2 disconnected parts and hole in one of them

Fig. 20 The four different effects of a cut

Any surface can be incised at any of its points, and cut along any of its curves. It will create two diffeomorphic curves that will be called the *lips* of the cutting. They belong to the boundary of the surface and from this fact are singular parts.

After cutting of a surface along one of its curve, one of the four following situations may happen (Fig. 20):

Such a cut may separate the surface into disconnected pieces. It happens each time the curve is the boundary of one or several disks on the surface (case 2 and case 4 where the surface has a curve as boundary, the curve along which the cutting is done meets that boundary into two points, creating with the boundary at least one loop which is the boundary of a 2-disk).

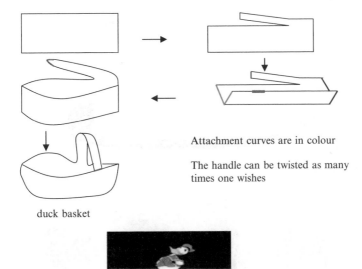

Attachment curves are in colour

The handle can be twisted as many
times one wishes

duck basket

Fig. 21 Duck like Donald

3 External Modifications

The cutting process, which may introduce separation into pieces, makes a transition
between internal and external modifications. If a separation has been introduced
through cutting, conversely, it is supposed that the inverse operation of joining the
two primitive separated pieces is possible. A good glue is all we need.

A *gluing process*, also named an *attachment process*, is considered here
as a change on the objects that are attached from the exterior side.

Such an addition is built along *domains of attachment*: it can be a point, a piece
of line, a piece of surface. The addition process supposes that the two objects that
will be attached share a similar domain which will be used as domain of attachment.

It should be noted that the process of adjunction can also work internally.

A simple example is the creation of a basket from a rectangle, involving cutting,
folding, and attachment. There are several ways to build this basket. Here is an
example (Fig. 21):

When the two domains of attachment belong to the same object, the attachment
process will be here called an *identification*.

The Möbius band is a classical result of an identification process. Take a band of
paper whose shape is a rectangle. Orient the two small edges in opposite directions,
twist the band an odd number of times, then you can glue (identify) the two small
edges since they have now the same orientation (Fig. 22).

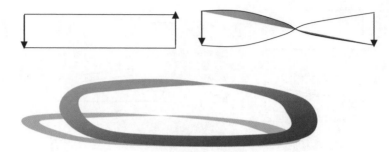

Fig. 22 The standard Möbius band

Fig. 23 Jeener's games with surfer

4 Synthesis

All the objects can be constructed using the following operations in the preceding sense: pinching, inflating, folding, cutting, attaching.

Here are some classical mathematical objects that can be built in that ways. They were sent to me by Patrice Jeener and produced with the software "surfer" (Fig. 23).

These objects were created by solving only one polynomial equation we will write under the abridged form $p(x_m, y_n, z_p) = 0$. Though for one polynomial equations the different classes of singularities are limited, one can expect an infinite number of shapes since the values of the integers m, n and p are themselves infinite: the human imagination cannot a priori reach the totality of the variation and subtle differences that exist between those mathematical shapes, all the more we consider here only one dimensional polynomials, whereas we can for instance consider projections onto 2- or 3-dimensional spaces of objects defined by various types of equations in multidimensional spaces. Except for a very few, like the ball or the cylinder, most of those mathematical objects do not have any significance to us, at this time: they are unfamiliar and considered as artificial; having no meaning, they seem to be cold and lifeless. But we cannot foresee the future. Humanity is evolving. Subjective interpretation may be giving way to more effective rational

Fig. 24 Two sculptures by Xavier Bonnet-Eymard

Fig. 25 Repetition in nature

thinking. Those objects may get a greater interest because they speak to our rationality through an intellectual training that teaches us how to look at them, how to see their properties and qualities. Though they may look richer by increasing their internal symmetries, each of these mathematical objects presented individually carries some level of melancholy due in part to their isolation.

Many artworks do not consist in the presentation of a single object. It happens of course: in that cases, the object has a sufficient strength of expression and richness *in se*, and sometimes appears as a composition of various objects. Sculpture, where the qualities of the material play an important role, is typical from this perspective (Fig. 24).

Artists rather create compositions. Several standard components play a role in those creations such as light, slightly distorted symmetry, abundance (mainly by repetition), perspective (from classical to reverse or frontal as in many Chagall' works) (Fig. 25). All these elements are related to physical fundamental principles and facts.

Maybe children, young and old ("Heraclitus called children' games men' thoughts"), will enjoy playing with some mathematical objects such as the ones we discussed before. They will be able to build friezes and free standing objects, fill their space with new creations, cut new shapes, create and hold new flowers, make new connections "à la Chagall" by inserting various objects and material in new composition. In that way, mathematics will be, as before, at the service of art.

Acknowledgments I am deeply indebted to Jean Constant who gave up so much time to correct all my Franglish ESMA papers, including that one. Many thanks also to Dick Palais and Simon Salamon and an anonymous author who did the same work on the text of the Preface.

References

1. Bruter, C.P.: Deux Universaux de la Décoration http://math-art.eu/pdfs/ConferenceSaverne.pdf (2010). Accessed 10 April 2010
2. Faber, E., Hauser, H.: Today's menu: Geometry and resolution of singular algebraic surfaces. Bull. Am. Math. Soc. **47**(3), 373–417 (2010)
3. Francis, G.K.: A Topological Picturebook. Springer, New York (1987)

Computer, Mathematics and Art

Jean-François Colonna

Abstract Mathematics could be seen as a "simple" mind game hardly more useful in everyday life than the chess. But their "formidable efficiency" as the language with which are written the laws of Nature could be an evidence they are the Reality. Thus, Mathematics would contain all works of art past, present and future, but also their creators.

1 Chess and Math

Prior to study concretely the links between Mathematics and Art, it must be first recalled what they are. Purists define them as a set of abstract symbols, manipulation rules and axiomatic statements regarded as true (and obvious) out of which are demonstrated theorems (that is to say new true statements). The progress (that is to say, the accumulation of new truths) are made in general thanks to problems posed to the community of mathematicians by one of its members. Most of the time, if not always, these issues are abstract and often incomprehensible to ordinary mortals and without apparent connection with reality. A recent example, which made the headlines, was the demonstration of Fermat's Last Theorem, completed in 1994 by Andrew Wiles after more than three centuries of trials and errors. And despite the immensity of the undertaking and the success met, here is a result of little practical usefulness[1] although the tortuous path for obtaining it was incredibly rich in developments and chance encounters. As presented, Mathematics would then seem to be a "simple" mind game hardly more useful in everyday life than the chess. Yet for

---

[1] The theorem itself (i.e., that the Diophantine equation $X^n + Y^n = Z^n$ has no solution for $n > 2$) has virtually no application, unlike some tools used for his demonstration; for example new methods of encryption using elliptic curves have emerged.

J.-F. Colonna (✉)
Centre de Mathématiques Appliquées, Ecole Polytechnique, Palaiseau, France
www.lactamme.polytechnique.fr

C. Bruter (ed.), *Mathematics and Modern Art*, Springer Proceedings in Mathematics 18, 47
DOI 10.1007/978-3-642-24497-1_5, © Springer-Verlag Berlin Heidelberg 2012

2,000 years and still more from the seventeenth century with Galileo, Mathematics are regarded as the language with which are written the laws of Nature. They are then, next to the microscope and the telescope, a revolutionary observation instrument who reveals to us every day new and mysterious aspects of our Universe.

2   Mathematics are the Reality

But these successes only make more mysterious their profound nature and their "formidable efficiency" (Eugene Wigner). What are Mathematics? Two seemingly irreconcilable answers can be formulated: either they are "only" the fruit of our minds, or they exist independently of us. Say differently: Is the mathematician like Molière who devised Monsieur Jourdain in Le Bourgeois Gentilhomme, or as Christopher Columbus who discovered America? Is the mathematician a creator (that is to say the one that pulls out of nothing) or an explorator (the one that travels observing)? Let us examine these two positions.[2] In the first case, Mathematics (our Mathematics!) are the brainchild of mathematicians;[3] they can be seen then as a language intended for the compression of the regularities observed in the nature.[4] In the second case, Mathematics are independent of us, so they exist outside of our time and our space, but where do they reside and what are they made of? This question seems quite puzzling, even crippling, but no more than to know where our universe is and what is it made of! It is even possible to go further (that is my point of view) and to consider that our Reality is a mathematical structure (among an infinite number of others) inside of which self-conscious sub-structures have emerged (us!). And so everything is "simple": Mathematics describe well the Reality because the latter is mathematical. So far two approaches complementary and seemingly disjointed, those of Art and Science,[5] were necessary to his knowledge; but if Reality is so, then the boundary blurs, disappearing and our computers are unaware[6] sub-structures that help us to gradually lift a "corner of the veil".

3   Art and Computer

But before examining the consequences of all this, let's describe briefly the concrete and pragmatic roles of the computer and Mathematics in the artistic field.[7] Appeared in the 1940s, the computer was quickly found in a strong position in all human activities

---

[2] In both cases, the mathematician is often (always?) doing Physics without knowing it.

[3] Matching could then be that the cortical structures of the senses (especially the sight) and of the "creation" are of the same nature and communicate with each other.

[4] As JPEG allows the compression of pictures.

[5] Respectively subjective and objective.

[6] But for how long?

[7] We will ignore here all that happened before the Second World War. Therefore it will be made no allusion to the golden ratio or again to the perspective.

and the constant progress in hardware and software areas have only amplified this phenomenon. This is especially true for industry, research or everyday life,[8] but it is much less in the Art world. From the beginning of the 1970s, I designed and developed the SMC (Conversational Multimedia System) system, a priori a system for computer aided instruction,[9] who was one of the first to allow both the synthesis and the processing of pictures (see Fig. 1). I realized very quickly the potential of digital techniques in audio and visual fields. In particular, all I announced more than 20 years ago at the conclusion of an article about the cinema aided by computer[10] is today a

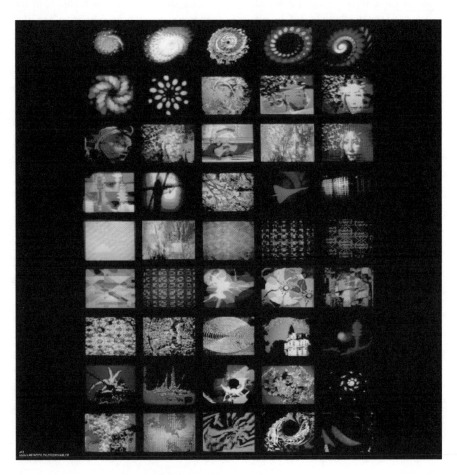

Fig. 1 Among my first coloured pictures (1974)

---

[8] What would everyday life be without the Internet?

[9] It was based on a T1600 Telemecanique computer with a memory of only 32 kilobytes while occupying several cubic meters!

[10] Colonna J.F. (09/1987) La Recherche, Paris.

reality in particular this convergence now taking place between movies and video games: the viewer has the opportunity to become an actor, while being able to immerse himself in realms unimaginable yesterday. But in the world of visual arts, the evolution was not as striking, far from it. Perhaps it is bound to certain questions: Who is the author? Where is the work of art? What is it made of? Is it perennial? Can we protect it? Is the computer able to create? And these questions are justified: indeed, the creation of a work of art by means of computers, next to the creative act itself, make use of softwares and hardwares that cannot be neutral.[11] And the work of art, is it what appears on a screen[12] or is it the string of binary digits which represents it in the memory of the machine or is it the set of comnands and gestures used to bring it to life? Moreover is it perennial: will it cross the chasms of time and will it be "readable" in several thousands of years as easily as the frescoes of Lascaux?[13] One of the

Fig. 2 Intertwinings and more (see www.lactamme.polytechnique.fr/Mosaic/descripteurs/ EntrelacsIntertwinings.01.Ang.html for more informations)

---

[11] For example, some features of a program can inspire an artist, allowing him something he would not have imagined without this intellectual prosthesis. Must the authors of these tools share the authorship of the work of art? This is the answer to this question that has frustrated most of my collaborations with traditional artists.

[12] Or on any media: paper or other.

[13] This question relates both to coding standards (like JPEG) as well as to media-recording devices such as DVD. To ignore it or to do nothing, implies the disappearance over time of an important part of our artistic and scientific heritage!

Fig. 3 Some fractal structures (see www.lactamme.polytechnique.fr/Mosaic/descripteurs/
AnimFractal.01.html and www.lactamme.polytechnique.fr/Mosaic/descripteurs/NDimensional-
DeterministicFractalSets.01.Ang.html for more informations

consequence of the principle of the digital representation of information is that copies
cannot be distinguished from the original: this means that on one hand the works here
loses its property of uniqueness and on other hand it can be duplicated more or less
easily,[14] with or without the consent of its author. Finally, the question of whether the
machine replace or will replace the artist is certainly the most delicate. From today, the

---

[14] There are ways of protection, such as watermarking, but I'm not convinced of their effectiveness
and especially their neutrality towards the work of art.

complexity of procedures that can be programmed may cause surprises (see Fig. 2), but obviously the machines lack the will to create and the awareness of themselves, but it is perhaps only a matter of complexity and therefore of time.

4   Art and Mathematics

I have solved myself most of these problems with the concept of potential work of art. In this context, the work of art is simply the underlying mathematical model. Obviously, this excludes from the creative field the works of art born from the artist's gesture, but this is not a limit for me, given the definition that was given previously of Mathematics: if they are indeed the Reality, they describe it in full! Let's illustrate this with an example: the one of the fractal geometry (see Fig. 3). Born in the 1960s with the work of Benoit Mandelbrot, it allows in particular to describe mathematically irregular and disordered phenomena unreachable using the Euclidean geometry. In this context, I developed a model that can produce images of cloudy mountainous landscapes, but also to animate them: the Alps as well as the American deserts or again the Moon. But all these images taken individually are not, in this context, the works of art and moreover it is impossible to display them all (and even more difficult to choose among them), since their number is strictly astronomical; the works of art are these equations that contain potentially all these images.[15]

Thus, Mathematics would contain all works of art past, present and future, but also their creators. Everything would be "written", but it doesn't matter since we have the consciousness to exist, to be free and to create.

---

[15] This design would certainly have pleased Jorge Luis Borges. He could make a sequel to his Library of Babel.

Structure of Visualization and Symmetry in Iterated Function Systems

Jean Constant

Abstract Principles and practices of visualization have always been valuable tools in all fields of research. The formation of representations plays a key role in all aspect of science. Prior research in the area of visualization demonstrates that representation of data hold great potential for enhancing comprehension of abstract concepts and greatly benefit collaborative decision-making and project performance. The purpose of this study is to examine the effectiveness of integrating various methodologies in the production of a visually coherent proposition. The author explores the dynamic of visualization in the digital environment and brings together elements of fractal topology, optical distortions and color theory components in an esthetic statement. The background and information selected for this purpose is based on specific cognitive processes, neurological and topological researches developed at the turn of the twentieth century by various experts in the field of investigative science: mathematicians Cantor and Sierpinski, neuro-scientists Hermann, Hering, Wundt and other analytical minds such as Itten and Kandinsky that, at a particular time in history, demonstrated the similarities and overall cohesion of an intellectual and scientific discourse that helped articulate the technological and esthetic world we know today.

1 Introduction

Georg Cantor was a prominent mathematician who mapped out a fundamental theory in mathematics at the turn of the twentieth century. He established the importance of one-to-one correspondence between sets. The Cantor set theory is particularly well suited for visualizing and experimenting with mathematical topics

J. Constant
Avenue Florimont 5, #3 1820 Montreux
e-mail: jconstant@hermay.org

C. Bruter (ed.), *Mathematics and Modern Art*, Springer Proceedings in Mathematics 18,   53
DOI 10.1007/978-3-642-24497-1_6, © Springer-Verlag Berlin Heidelberg 2012

Fig. 1 Cantor deconstructed

such as iterated-function systems, similarity dimension, complex numbers, con-
nectedness, and topological groups (Fig. 1).

Topology is a major area of mathematics concerned with spatial properties.
It emerged through the development of concepts such as space, dimension, and
transformation. An iterated function system can be defined as an attractor formed
by the union of a finite number of contraction or expansion mappings.

Some well-known generalizations of the Cantor set can be found in the
Hausdorff's dimension similarity principle and the Sierpinski's square and triangle
named after its author.

Sierpinski's iteration occurs when an object or image is transformed by a set of
affine alterations to produce a new image. The new image is then modified by the
same affine conversion to produce another new image.

At the particular time (mid 1800–mid 1900) mathematicians were mapping
those various theories, authorities in other disciplines such as psychological and
physiological sciences were investigating optical aberration. Ewald Hering,
Ludimar Hermann produced templates that had uncanny visual likeness to the
work of Cantor or Sierpinski. Also to be noted, this time also saw the reevaluation
of traditional 2 dimensional art paradigms with the arrival of cubism,

constructivism and the many movements that lead to a radical reinterpretation of the esthetic environment. All elements very relevant to the object of this study.

The author's background being into fine art, and visual communication, the reader needs to be made aware that precision and exactitude in the field of mathematic may be at time only tentative. For more scientific rigor on fractal and chaos theory, one is encouraged to explore the work of more qualified experts in Laplacian mathematic, the papers of Pelander [1] and Barlow and Bass [2] on the solvability of differential equations of open subsets of the Sierpinski gasket, and other professional publications in the field of cognitive and biopsychology science.

2   Methodology

This experiment evolved around the following three elements:

1. A mathematical part to revisit fundamental concepts of set theory, symmetry and recursive function in topological spaces.
2. A technical part to develop graphic visualizations in a digital environment.
3. A subjective and artistic part to infuse various esthetic interpretations with the help of alternate color propositions.

2.1   Data Collection: Mathematical Iteration Systems

According to Falconer [3], one of the essential features of a fractal is that its Hausdorff dimension generically strictly exceeds its topological dimension. Because they appear similar at all levels of magnification, fractals are often considered to be infinitely complex. Mandelbrot [4] and Julia sets offer today extraordinary complex visualizations made possible through the expression of computerized technology. However it is to be noted that the outcome, while technically sound, often ends up defeating a major component of visual statements because of the mechanical and repetitive nature of its expression: the esthetic environment seems to require for most some level of uncertainty well within the tissue of what contributes to human perception of reality.

Waclaw Sierpinski was the mathematician who at the turn of the century mapped visually the Cantor set theory in simple visual terms. He first became interested in set theory when he came across a theorem that stated that points in the plane could be specified with a single coordinate. Two popular visualizations that emerged from his research are known today as the Sierpinski gasket (triangle) and the Sierpinski carpet (square). A key element that made it a natural choice for this study on the effectiveness of a message initiated on a flat 2D surface is the fact that the Sierpinski iteration originates and progresses at a straight angle where most commonly rendered fractals are based on diagonal and curve progression.

Fig. 2 Starting grid

2.2 Graphic Manipulation

Vector graphics store lines and shapes as mathematical formulae. They provide accurate visual rendition and produce an image scalable to any size and detail. All compelling factors when keeping open multiprocessing output options and transfer mathematical information with exactitude. In the words of Dr. Konrad Polthier [5] "Mathematical visualization driven by computer graphics has proven to be a successful tool for mathematicians to investigate difficult mathematical problems".

2.2.1 Vector Outline

Adobe Illustrator is a sophisticated vector based graphic software. It allows the user to explore and visualize mathematical relationships in terms of 2D line and precision drawing. It was selected as the best available tool to translate specific mathematical concept in visual terms and in particular for a system that transforms a sequence of symbols into a unique set of points in the 2-dimensional space. The Sierpinski square pattern requires that one starts with a solid square and divide it into nine smaller congruent squares. The original square is scaled 8 times by a factor 1/3. Measurement standards were adjusted to the metric system to make calculation easier.

To solve the issue afferent to the identification of the squares in the progression, a grid 20% black down to 80%, was created and filled accordingly, starting with the 1 mm^2(80% black) down to 20% for the eighth iteration (Fig. 2).

Cartesian coordinates and topological dimensions defined by Sierpinski's contemporaries Lebesgue (1903) and Hausdorff (1918) set the original measurement (point in space) as zero. For all practical purpose, this was not viable information to start from in the particular environment in which the demonstration was developed. A unit of one was chosen as the starting point. The total virtual canvas size would be for instance 2,187 mm, a size one assumed the processing power of the computer could handle.

Computer profile: Hardware:
Model Name: Mac Pro
Model Identifier: MacPro 3.1

Processor Name: Quad-Core Intel Xeon
Processor Speed: 3.2 GHz
Number Of Processors: 2
Total Number Of Cores: 8
L2 Cache (per processor): 12 MB
Memory: 8 GB
Bus Speed: 1.6 GHz
Boot ROM Version: MP31.006C.B05
SMC Version: 1.25f4
Graphic & Display
NVIDIA GeForce 8800 GT:
VRAM (Total): 512 MB
Software:
– Mac OSX v.10.5.8
– Adobe CS4 Photoshop, Illustrator, Bridge
– Adobe-Lab Kuler API (Application Programming Interface)
– Keynote 09
– QuickTime Pro

Moving to plate 243 mm, slight changes of x y/w h positioning started to be noticed and dimension of the smaller squares, (600.0007 mm turned into 600.009 mm in a simple copy/paste step) even after locking all elements. The computer system started to slow considerably during rendering. Also it was noted that the center square of plate 243 (Figs. 3a, b) filled with 40% black disappeared in the background because of the hue interference of too many squares 1, 3 and 9.

To be consistent with the premises of the demonstration and bring the progression to 2,187, it was decided to proceed with a reverse elimination of square-layers 1, 3, and 9 mm, except in the larger squares (Fig. 4).

The end result shows intriguing optical similarities with symmetry patterns and tiling mentioned in CP Bruter expose in Saverne [6], and that were developed over centuries in various civilizations across the world (Fig. 5).

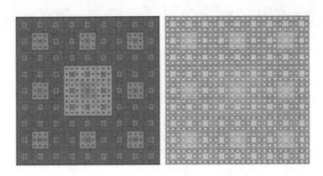

Fig. 3a Plate 243

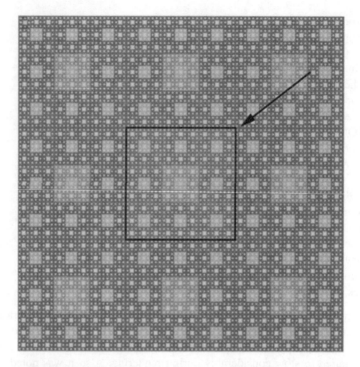

Fig. 3b Central square first step

Fig. 4 Improving the central square

A fact made all the more prevalent by the monochrome tone of the visualization that brings up the design element of the statement (Fig. 6).

2.2.2 Optical Illusion

Can we really trust what our eyes see?

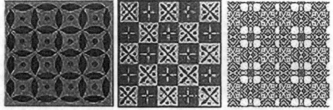

Egyption designs   H.Field

Fig. 5 Local patterns

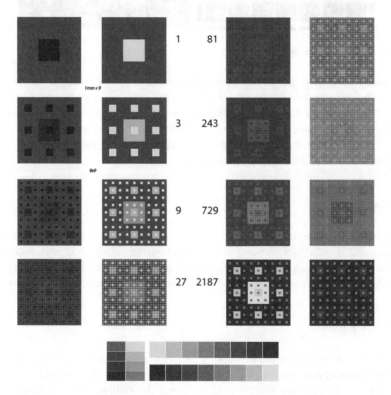

Fig. 6 Summary

An interesting property of an iteration in a two dimensional environment is that it is most likely to take the viewer off balance both for cultural and physiological reasons: anchor and points of reference are arbitrary and the determination of objects dynamic and elements of perspective idiosyncratic. The history of art is an ongoing experiment on the mapping of this particular challenge. It also outlines the extent and limitation of communicating on a flat surface. Science for the best part of the last 200 years has brought a deep, universal, objective understanding to intuitive exploration of this universe made over centuries by talented artists in various cultures, worldwide [7].

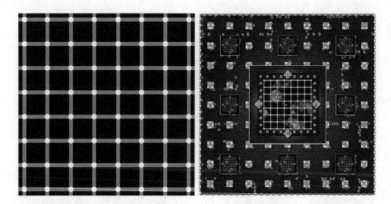

Fig. 7 Using the Hermann grid

In the following stage, we attempted to reconcile and experiment in simple terms the findings of various disciplines involved in the exploration of human perception as well as the more subjective element of esthetic interpretation. Plate 2187 was transferred in a graphic editing program (Photoshop 10.00) where selected elements were extracted, blended, and recomposed as needed in new visualizations.

Below is a brief descriptive of the process

Hermann Grid

Dark patches appear in the street crossings, except the ones that you are directly looking at. If you look around in the neighboring figure you will notice the appearance and disappearance of black dots at the crossings (Invoked to explain Florida's election problems in 2001: "Count the black dots, recount to confirm. . .").

The Hering Illusion

Discovered in the nineteenth century by German physiologist Ewald Hering. Two straight lines appear curved or bowed in the context of intersecting lines with orientations that change progressively. Hiding the oblique lines from the view will reveal the fact that the horizontal bars are equally straight (Fig. 8).

The Wundt Illusion

Considered the founder of experimental psychology, Wilhelm Wundt introduced cognitive principles to the psychology community in the late 1880s. He was the first one to report on the phenomenon. The overestimation of the sub tense of acute angles and the underestimation of obtuse ones was first reported first by Wundt (Fig. 9).

Fig. 8 Using the Hering illusion

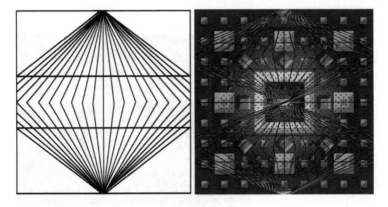

Fig. 9 Using the Wundt illusion

The Aitken Wheel

In homage to Robert Grant Aitken astronomer, director of the Lick observatory, president of the University of California. Minor planet (3070) Aitken is named in his honor as well as a Moon crater 2 on the far side.

From another Aitken: Concentrating on the wheel center for a few minutes will make the wheel spin (Fig. 10).

The Ebbinghaus

Identical objects are perceived as unequal in size when objects of a different size surround them. This kind of assimilation illusion is also associated with the name of Franz Joseph Delboeuf, a Belgian philosopher. The copies of the same shape—circles and rings in our case—appear to inherit properties of their environment.

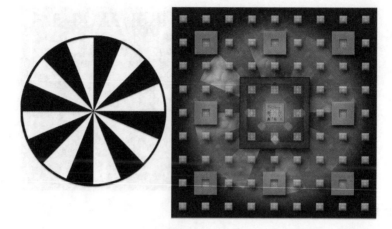

Fig. 10 Using the Aitken wheel

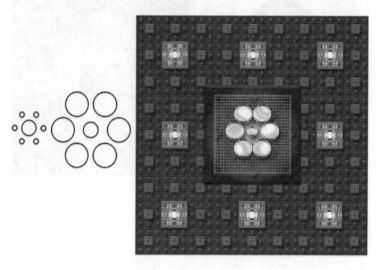

Fig. 11 Using the Delboeuf illusion

Two equal circles—one inside a bigger circle, the other containing a smaller circle—seem to have different sizes (Fig. 11).

2.2.3 Color Scheme

Up to this point, the demonstration had been conducted strictly in a terms of black and white and shade of gray. Grayscale images allow to concentrate on measuring the intensity of light and the electromagnetic spectrum of any given frequency.

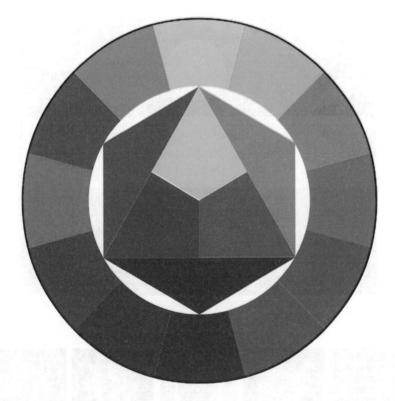

Fig. 12 Itten's color wheel

It helps focus in precision mapping and objective rationalization of effect for best possible outcome.

In the early 1900s, Johannes Itten [8], color instructor at the Bauhaus and trained in psychoanalytic theory developed a color wheel that took into consideration the subjective feeling that's associated with objective color, and the psychic and emotional values of colors (Fig. 6). Itten's color wheel was used as a reference to expand this project and include another layer of visual interpretation (Fig. 12).

This work was made considerably easier by a very dynamic color tool from the Adobe Lab, the Kuler application (Fig. 7). Its algorithm allows the user to create and test color combinations in real time and download them on the workspace as needed (Fig. 13).

Below is an example of the progression from black and white to color (Figs. 14 and 15):

2.3 Dissemination

This part of the expose is relatively succinct for a simple reason: the technological context in which we communicate is fast changing. Specific mechanical

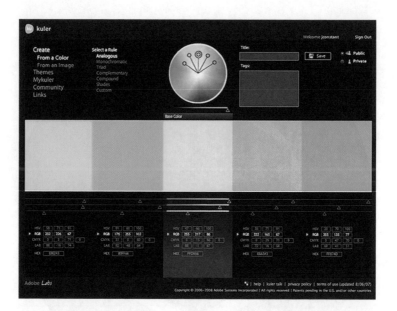

Fig. 13 The Kuler application

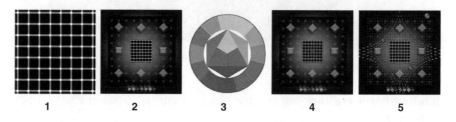

Fig. 14 From the grid to the application

information that may be appropriate today could fast become obsolete in tomorrow environment.

However one compelling element legitimates mention of this stage of production: works of art as well as works of science are meant to be shared. Whatever the technology at hand, the ultimate goal of any endeavor of that nature is to be shared and benefit as many as possible so they can learn from the many blunders and missteps and hopefully build from some of the inspiration it provides.

2.3.1 Multimedia

The goal of developing a multimedia presentation was dual: summarize the experiment in an attractive presentation and use various techniques to enhance key points of the demonstration in visual terms. The final product was encoded for QuickTime

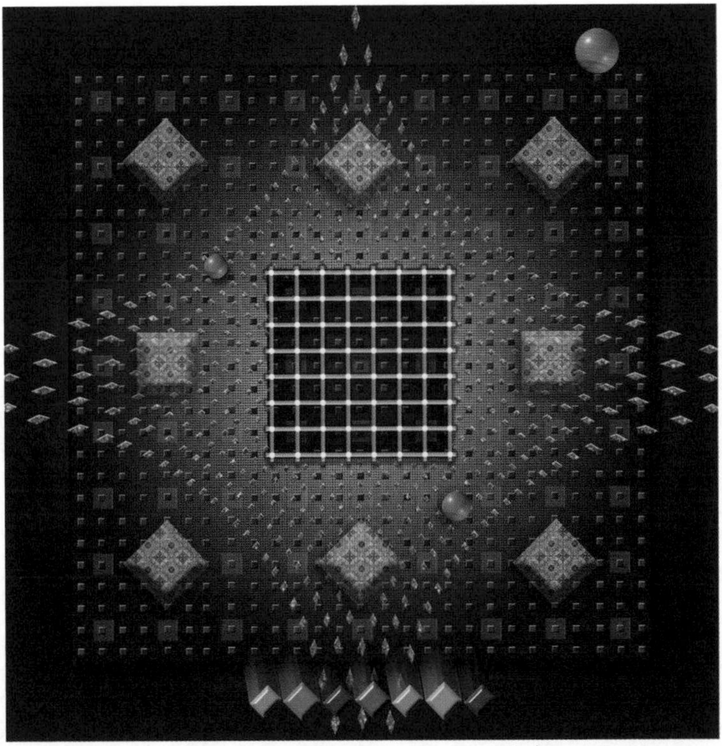

Fig. 15 Final result

viewing—QuickTime being a media viewing platform available on most computers today.

Visual Component

The Apple Keynote interface is among the most prevalent forms of persuasive technology used today in all fields of communication. Its powerful engine and multiple effects allow authors to maximize their effort and reach out effectively targeted audience. The progression of the experiment was extracted in a series of single slides that were combined with appropriate transitions and effects available in the program toolbox.

Audio Track

Professional animators believe that 70% of a moving-images presentation impact comes from the sound track. Sound guides the rhythm of the presentation; helps emphasize specific part and keep the viewer attention focused in the progression. The soundtrack background was provided by a Mikael Kott, a classically trained artist and remixed in an audio software to fit the particular of the presentation.

2.3.2 Deployment

The emergence of the Internet and other tools of modern communication have radically changed the ways by which one evaluates how to present a body of work. Twenty one original plates were printed on fine art, archival paper, and size 22×22 inches to meet more classic method of diffusion. The digital multimedia presentation was made available on CD and uploaded both on the author's website and YouTube, a highly visible Internet social media platform. The h.264 encoding of the final presentation also makes it available for viewing on IPods and mobile devices.

3 Conclusion

As it happens often in exploratory journey, I found challenges I was not expecting and got rewards I was not planning on finding. It helped me develop a deeper respect for the mathematic and scientific methodologies and the minds that brought us objective understanding of our physical environment. It also expanded my appreciation of the engineers and programmers that allow us today to expand the understanding of abstract visualizations in such complex terms. Computerized technology has made research, in particular in the field of iterated systems, much more dynamic.

The coherence of scientific and artistic concepts at the turn of the past century still stand today as solid technical and intellectual propositions all fields of communication can benefit from. The tools of mathematics, psychological and optical physiology enhance visual propositions in all forms of expression and expand its effectiveness.

Where do we go from there? [9]

This quick overview of concepts shared across disciplines by experts demonstrates again that, if our operating systems are neither finite nor perfect, the ongoing collaboration of trained specialists in various disciplines continues to ensure that tomorrow will be brighter as it opens the door to new advancement in the understanding and appreciation of our circumstances.

Acknowledgments I would like to express my deep and sincere gratitude to the following: Dr. Anders Pelander, Dr. Ruby Lathon, C.P. Bruter, Ulrich Niemeyer, Roland Piquepaille, Lokesh Dhakar, Alessandro Fulciniti, Brigitte Furst, Steve Wold and the many other who provided me with inspiration, advice and technical guidance.

References

1. Pelander, A.: A Study of Smooth Functions and Differential Equations on Fractals. Department of Mathematic. University of Upsalla, Sweden (2007)
2. Barlow, M.T., Bass, R.F.: The construction of Brownian motion on the Sierpinski's carpet. Ann. Inst. Henri Poincaré **25**(3), 225–257 (1989)
3. Falconer, K.: Techniques in Fractal Geometry. Wiley, NY (1997)
4. Mandelbrot, B.: Fractals: Form, Chance and Dimension. Freeman and Company, CA (1977)
5. Polthier, K.: Visualizing Mathematics. Technische Universität, Berlin (2002)
6. Bruter, C.P.: Deux Universaux de la décoration. Conférence Saverne. http://math-art.eu/pdfs/ConferenceSaverne.pdf
7. Albright, T.D., van der Smagt, M.J.: Motion after-effect direction depends on depth ordering in the test pattern. Vision Center Laboratory & Howard Hughes Medical Institute, The Salk Institute for Biological Studies (2001)
8. Itten, J.: The Elements of Color. Wiley, NY (1970)
9. Palais, R.: The visualization of mathematics: Towards a Mathematical Exploratorium, Notices AMS **46**(6), 647–658 (1999)

M.C. Escher's Use of the Poincaré Models
of Hyperbolic Geometry

Douglas Dunham

Abstract The artist M.C. Escher was the first artist to create patterns in the hyperbolic plane. He used both the Poincaré disk model and the Poincaré half-plane model of hyperbolic geometry. We discuss some of the theory of hyperbolic patterns and show Escher-inspired designs in both of these models.

1 Introduction

The Dutch artist M.C. Escher was known for his geometric art and for repeating patterns in particular. Escher created a few designs that could be interpreted as patterns in hyperbolic geometry. Figure 1 is rendition of Escher's best known hyperbolic pattern, *Circle Limit III*. Escher created his hyperbolic patterns by hand, which was a very tedious and time consuming process, since the motifs were of different sizes and slightly different shapes. So about 30 years ago my students and I were inspired to create such patterns using a computer, which could transform the motifs almost instantly. In this paper we show some of the hyperbolic patterns we have generated.

   We begin with a brief history of the creation of artistic hyperbolic patterns. Then we review the Poincaré models of hyperbolic geometry, and repeating patterns. With that background, we next show sample patterns from both the disk and half-plane models. Finally, we indicate possible directions of further research.

2 A Brief History of Hyperbolic Art

Euclidean, spherical (or elliptic), and hyperbolic geometry are sometimes called the "classical geometries". The Euclidean plane and the 2-dimensional sphere are familiar since they can be embedded in the 3-dimensional space in which we live.

D. Dunham (✉)
Department of Computer Science, University of Minnesota, Duluth, MN, USA
e-mail: ddunham@d.umn.edu

C. Bruter (ed.), *Mathematics and Modern Art*, Springer Proceedings in Mathematics 18,   69
DOI 10.1007/978-3-642-24497-1_7, © Springer-Verlag Berlin Heidelberg 2012

Fig. 1 A computer rendition of the *Circle Limit III* pattern

However, there is no smooth isometric embedding of the hyperbolic plane in Euclidean 3-space, as proved by David Hilbert more than 100 years ago [5]. Thus we must rely on non-isometric models of it. This is probably the reason for the late discovery of hyperbolic geometry by Bolyai, Lobachevsky, and Gauss almost 200 years ago. And it wasn't until the late 1860s that Eugenio Beltrami discovered what are now called the Poincaŕe disk and half-plane models of the hyperbolic plane.

Almost a century later Escher received a copy of a paper from the Canadian mathematician H.S.M. Coxeter [1]. The paper contained the hyperbolic triangle pattern shown in Fig. 2. Escher said that the Fig. 2 pattern gave him "quite a shock" since it showed him how to make a repeating pattern with a circular limit (hence the name for his "Circle Limit" prints); he was already familiar with patterns with point limits (with dilation symmetries) and "line limits". Thus inspired, Escher created *Circle Limit I* in 1958, a rendition of which is shown in Fig. 3. Over the next 2 years Escher created three more "Circle Limit" prints:

Circle Limit II, Circle Limit III (shown Fig. 1), and *Circle Limit IV*. For more information, visit the official Escher web site [6]. Twenty years later, my students and I were in turn inspired to re-create Escher's four "Circle Limit" patterns using computer technology [3]. However the program we wrote was more general than required to reproduce Escher's "Circle Limit" patterns, so we created a number of new hyperbolic patterns. Another reference for the theory of computer generated hyperbolic patterns is [2].

Fig. 2 The 6,4 tessellation

Fig. 3 A *Circle Limit I*
rendition

3 Repeating Patterns and the Poincaré Disk and Half Plane Models

A model of hyperbolic geometry represents the basic elements of that geometry (points, lines) by Euclidean constructs. Conversely, as Beltrami showed, there are models of Euclidean geometry within hyperbolic geometry, so that two geometries are equally consistent.

In the *Poincaré disk model* of hyperbolic geometry the hyperbolic points are represented by Euclidean points within a bounding circle. Hyperbolic lines are represented by (Euclidean) circular arcs orthogonal to the bounding circle (including diameters). The edges of the triangles in Fig. 2 and the backbone lines of the fish in Fig. 3, are hyperbolic lines. However the backbone lines of the fish in Fig. 1 are not hyperbolic lines, but are so called equidistant curves (each point is the same distance from the hyperbolic line with the same endpoints on the bounding circle), which make an angle of about 80° with the bounding circle. The hyperbolic measure of an angle is the same as its Euclidean measure in the disk model—the model is *conformal*, so that motifs retain the same approximate shape as they approach the bounding circle. This was a property of the disk model that appealed to Escher. Another desirable property was that an entire pattern could be displayed in a finite area, unlike "point limit" patterns which could theoretically grow outward to infinity and, "line limit" patterns which could also extend to infinity upward and to the left and right. However, equal hyperbolic distances correspond to ever-smaller Euclidean distances toward the edge of the disk, thus all the fish in Fig. 1 are the same hyperbolic size, as are the triangles in Fig. 2.

In the *Poincaré half-plane model* of hyperbolic geometry the hyperbolic points are represented by Euclidean points (x, y) in the upper half plane y>0. Each hyperbolic line is represented by a (Euclidean) semicircular arc above the x-axis and with center on it (including vertical half-lines). Figures 4 and 5 show half-plane versions of Figs. 2 and 3 respectively. The edges of the triangles in Fig. 4 and the backbone lines of the fish in Fig. 5 are all hyperbolic lines in this model. This model is also conformal, but was not as appealing to Escher as the disk model since it is unbounded. Still, Escher used this model to create two and possibly three patterns, which he called "line limit" patterns. The hyperbolic distance relationship is simple in this model—hyperbolic length is inversely proportional to the Euclidean distance to the x-axis.

A *repeating pattern* is a pattern made up of congruent copies of a basic subpattern or *motif*, where "congruence" is determined by the geometry in question. In Fig. 1, the motif consists of one fish (disregarding color). In Figs. 2 and 4, the motif can be either a black or a white triangle (again disregarding color). The motifs of Fig. 3 and 4 consist of half a white fish together with an adjacent half of a black

Fig. 4 A half-plane version of the triangle pattern of Fig. 2

fish. It seems necessary to use repeating patterns to show the hyperbolic nature of the models. For instance, if there were just one triangle shown in Figs. 2 or 4, we couldn't be sure if it was hyperbolic or just a curvilinear Euclidean triangle. For more information on hyperbolic geometry and its models, see [4].

4   Patterns in the Poincaré Disk Model

For completeness, we show renditions of Escher's patterns *Circle Limit II* and *Circle Limit IV* in Figs. 6 and 7. Escher's last print, *Snakes* contains a pattern of interlocking "hyperbolic" rings near the circular boundary; the inner rings form a

Fig. 5 A half-plane version of Escher's *Circle Limit I* pattern

Fig. 6 A *Circle Limit II* rendition

Fig. 7 A *Circle Limit IV*
rendition

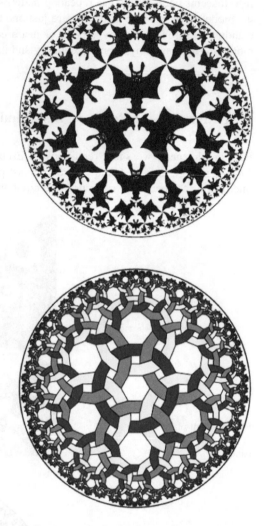

Fig. 8 A interlocking ring
pattern inspired by Escher's
Snakes

"point limit" (dilation) pattern. Figure 8 shows a complete pattern of the hyperbolic
rings. Figure 9 shows a pattern like *Circle Limit III*, but with five fish meeting a
right fin tips.

5   Patterns in the Poincaré Half-Plane Model

Escher seems to have created three "line limit" patterns. His *Regular Division of the
Plane VI*, Fig. 10, and *Square Limit* are based on the half-plane model, and *Regular
Division Drawing 101* may be, but it is hard to tell since the lizards are modified in

Fig. 9 A pattern of fish, five of which meet at right fins

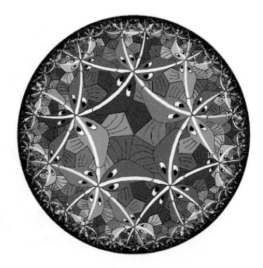

Fig. 10 A Escher "line limit" pattern

different ways. Figures 11–13 show half-plane versions of *Circle Limit IV*, the pattern of Fig. 9, and a fish pattern inspired by *Regular Division Drawing 20*.

Fig. 11 A half-plane version
of *Circle Limit IV*

Fig. 12 A half-plane version
of Fig. 9

6 Future Work

The disk model patterns we have created were designed using a drawing program
that works in that model. This program has evolved over the years to have a number
of useful features. However, the half-plane patterns that we have created were first
designed using the disk model program and then transformed to the half-plane
model. It would seem to be useful to have a program that would allow for the design
of half-plane patterns using that model directly.

Also, we have just shown a few patterns in each of the models. It would be
interesting to create many more such patterns.

Fig. 13 A half-plane fish pattern

Acknowledgments I want to thank Lisa Fitzpatrick, director, and the staff of the Visualization and Digital Imaging Lab at the University of Minnesota Duluth.

References

1. Coxeter, H.S.M.: Crystal symmetry and its generalizations. Trans. R. Soc. Can. **51**(3), 1–13 (1957)
2. Dunham, D.: Hyperbolic symmetry. Comp. Math. Appl. **12B**(1/2), 139–153 (1986); Also appears in the book Symmetry István, H. (ed.). Pergamon Press, New York (1986); ISBN 0-08-033986-7
3. Dunham, D., Lindgren, J., Witte, D.: Creating repeating hyperbolic patterns. Comp. Graph. **15** (3), 215–223 (1981); August (Proceedings of SIGGRAPH'81)
4. Greenberg, M.: Euclidean & Non-Euclidean Geometry: Development and History, 4th edn. W. H. Freeman, New York (2008); ISBN 0-7167-9948-0
5. Hilbert, D.: Uber Fläschen von konstanter gausscher Krümmung. Trans. Am. Math. Soc. **2**, 87–99 (1901)
6. Official Escher web site: http://www.mcescher.com/

Mathematics and Music Boxes

Vi Hart

Abstract Music boxes which play a paper tape are fantastic tools for visually demonstrating some of the mathematical concepts in musical structure. The literal written notes in a piece can be transformed physically through reflections and rotations, and then easily played on the music box. Principles of topology can be demonstrated by playing loops and Möbius strips. Written music can also be transformed into different types of canons by sending it through multiple music boxes.

1 Introduction

Music boxes are familiar objects to most people, but most music boxes are manufactured to play only a single song. Some may be able to play a set of "records" manufactured for it, but even these are made specifically to play only the intended input in the intended way. Still, there is another kind of music box that plays a paper tape with holes punched in it [1]. Not only can the user write their own music by punching their own paper strips, but there is enough freedom in how the strip is played that the music can be transformed. Transformation is a basic concept in both music theory and geometry. Two musical phrases, or two triangles, can be congruent even if they look or sound different. These music boxes allow one to see notes as a set of points punched in paper, making both mathematics and music a visual, auditory, and tactile experience.

V. Hart (✉)
7 Circle Dr. Farmingdale, NY 11735, USA
e-mail: vi@vihart.com

C. Bruter (ed.), *Mathematics and Modern Art*, Springer Proceedings in Mathematics 18,
DOI 10.1007/978-3-642-24497-1_8, © Springer-Verlag Berlin Heidelberg 2012

2   Related Work

Much has been written on the subject of music and mathematics, and the relationship between musical and mathematical transformations [2]. Others have also noted the mathematical potential of the music box, specifically with Möbius strips [3, 4].

3   Transforming a Piece

A blank strip of paper can be transformed four ways and still appear the same as it started. One could flip it vertically, flip it horizontally, or turn it upside-down, or of course leave it alone. Once it has holes punched in it, these transformations correspond to the transformations taught in every introductory music theory course: a motive (identity) can be found in inversion (reflection through horizontal axis), retrograde (reflection through vertical axis), or retrograde inversion (180° flip). Using a music box, these are the four ways one could insert the same strip of paper into the box.

Another way to think of these transformations is that the music, or points in space, are staying the same, but how we look at it is changing. One could think of the music box as a reader, scanning the strip in four different ways, as seen in Fig. 1.

The strip of paper can also be shifted forward or backwards in time, by playing it now or later. While this is not usually thought of as a transformation to a musician, it is known in geometry as a horizontal translation. If the strip is thinner than the width of the music box, the paper can also be translated vertically. This will make the pitch higher or lower, known to musicians as transposition. The two can be combined to play the music at both a different time and pitch, as seen in Fig. 2.

There is a difference between these musical transformations and transformations on the Euclidean plane.

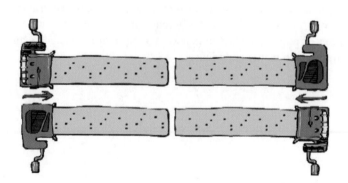

Fig. 1 Four ways to read the same strip of tape

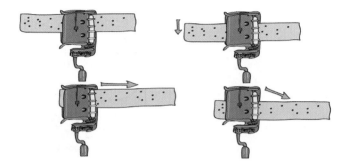

Fig. 2 Translating vertically, horizontally, or both

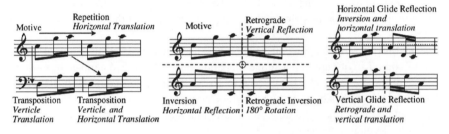

Fig. 3 Isometric transformations in musical space

The musical plane has two distinct axes: time and pitch space. The only transformations allowed are ones that do not confuse the two axes. It would not make sense, for example, to try to put the strip through sideways (one could cut it down to a square and physically do it, of course, but that sort of transformation artificially sets a conversion rate between space and time).

Because of this, some kinds of transformations have more than one version. A reflection along the vertical axis is different from a reflection along the horizontal axis, and playing a piece at a higher pitch is different from playing it tomorrow.

The basic transformations of reflections and translations can be combined together to get glide reflections. All together, there are eight distinct isometric musical transformations, as shown in Fig. 3 [5].

4   Loops

One can make the music repeat by physically taping the paper into a loop, after it has already been put through the music box. If you put a twist in the paper before taping it, you get a Möbius strip, as shown in Fig. 4. This gives you a pattern of the theme and its inversion. The music plays through once normally, and when it gets back to the beginning of the loop the paper will be taped on upside-down, and so the

Fig. 4 Möbius music box

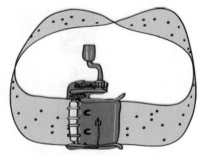

inversion will play. Putting two twists in the paper sounds the same as no twists, or indeed any even-number of twists. What matters is the topology of the loop: does it have one side or two? The music box only "scans" along one side, so if the paper has one side as a Möbius strip does, it will play both sides of the original un-twisted strip.

The Möbius strip is a fun construction because you get twice the length of music out of your strip of paper, in a single repeating form. If you had a music box where you could turn the handle backwards to go backwards, you could start playing it backwards to get a pattern of the retrograde and retrograde inversion. In the case of the available box, once you tape your paper in a loop you are stuck going one way only (until you untape it).

The exception would be a theme that is, in music theory terms, "uninvertable," meaning that it sounds the same under inversion, and would therefore have a horizontal line of mirror symmetry. If this line of symmetry is at the same pitch that the music box inverts, the theme would sound the same when put through the box upside-down. If it weren't, the inversion would transpose the music.

A video demo of a Möbius music box is available on my website [6].

5 Canons

By putting a strip of paper through more than one music box, as seen in Fig. 5, one can play a canon. The same music is played twice, with a time shift on the overlap depending on how far apart the boxes are. One can also think of this as two scanners reading a strip at the same time.

6 Types of Canons

A standard canon has only a shift in time between the two (or more) boxes, but there are many other types that use other transformations between the boxes. As shown in Fig. 6, one could also twist the piece of paper between the boxes to get another kind

Fig. 5 Music box canon

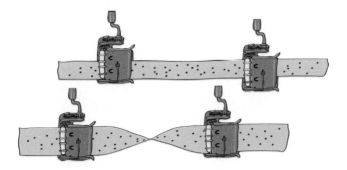

Fig. 6 More kinds of music box canons

of canon, where the theme is followed by the inversion, or put a vertical shift between boxes to get a canon at the 5th (or any other interval in range). There are also types of canons where one voice plays the music forwards and the other in retrograde, for a crab canon, or upside-down and backwards, for a table canon, though unfortunately in this case the music boxes would run into each other and get stuck.

A mensural canon is one where the voices play the same music at different tempos, which can be done by playing the boxes at different speeds. Some of these types of canons may not be as well known to nonmusicians, but there are many examples by Bach, Mozart, and others.

Some of Bach's canons, notably in his *Musical Offering*, were presented as puzzles, with the music of the theme written out only once instead of showing what each instrument plays at each time. The music had clues as to which transformation would yield a good sounding piece, and it was up to the musicians to puzzle it out. Music boxes allow one to easily try out different combinations.

One of the best uses of music box canons is how it makes it easy for non-musicians to see and understand the structure of a canon. I demonstrate this in a video too, using a popular and easily recognized example: Pachelbel's Canon. Three music boxes are used to represent the three violins, and all three physically play the same strip of paper, one after another. The bass line, or basso continuo, is played by a fourth music box, playing a tight loop of paper that shows the repeating nature of the bass line. The form of the piece becomes easy to see [6].

References

1. Kikkerland "DIY Mechanical Music Box Set:" http://www.kikkerlandshop.com/toys-games-musicboxes.html
2. Hodges, W.: The geometry of music. Music and Mathematics: From Pythagoras to Fractals, pp. 91–111. Oxford University Press, London (2003)
3. Ranjit Bhatnagar's Möbius music box photo: http://www.flickr.com/photos/ranjit/3314000751/
4. Tremblay, R.: Inversus: Sankyo 20-Note Moebius Strip Plays Inverse Music, Mechanical Music Digest (2003). http://www.mmdigest.com/Sounds/Sankyo20/tremblay.html
5. Hart, V.: Symmetry and transformations in the musical plane. In: Proceedings of the 12th Annual BRIDGES Conference: Mathematics, Music, Art, Architecture, Culture (BRIDGES 2009), Banff, Canada (2009)
6. Hart, V.: Music Boxes: http://vihart.com/musicbox/

My Mathematical Engravings

Patrice Jeener

Abstract The work of an engraver is shown through the presentation of three types of engravings concerning minimal surfaces, closed surfaces without singularities, and bi-periodic functions.

1 Introduction: The Engraver's Job

I make engravings on copper. From the following images, one can follow some of the many steps that end in the production of an image from the engraving. In this case it is an engraving of an olive tree that could be used as an emblem for Provence, the region where I live and which I try to honor with some of my works (Figs. 1–6).

Engraving: The engraving method consists of cutting incisions on a copper plate with a kind of chisel called a burin. The finished engraving is then printed using the intaglio method. That is, the entire surface of the plate is coated with ink and, after wiping, the ink remains in the incised markings. The printing press consists of two assembled steel cylinders, one atop the other. The inked copper plate is then put on a steel plate on which there is placed a wet piece of Arches paper and a felt cloth; all of which is then passed between the cylinders. The result gives a light relief of the incisions on the paper.

My main inspiration, however, is different. It is in mathematics that helped me discover the meaning of models shown at the Institute Henri Poincaré, and it reminds me on the time of my youth, at the Palais de la Découverte (see the article by Francis Apéry). If I take each of my visits to our capital to make prints of the rich Parisian scenery, it is in Provence that I compose my engravings of mathematics. I will now give an overview on three themes, namely, minimal surfaces, closed surfaces without singularity, and bi-periodic functions.

P. Jeener
27 Grande Rue, 26470 La Motte-Chalancon, France
e-mail: patrice.jeener@wanadoo.fr

C. Bruter (ed.), *Mathematics and Modern Art*, Springer Proceedings in Mathematics 18, DOI 10.1007/978-3-642-24497-1_9, © Springer-Verlag Berlin Heidelberg 2012

Fig. 1 Engraving

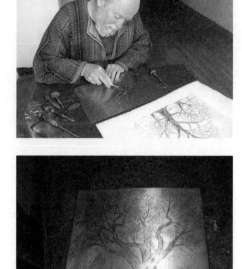

Fig. 2

Fig. 3

2 Minimal Surfaces

I started by studying certain surfaces of 3rd and 4th degree, given by simple equations that I could solve easily. I executed then lines thanks to the Cavaliere perspective, drawing carefully each of their remarkable curves. When data processing appeared, I programmed myself in BASIC to compute these surfaces starting from their parametric equations—today, I use standard software solutions. It was enough for me to choose the contour and the visual angle of my surfaces. Starting from a screen printing, I obtained a copy of a basis model being used for the future engraving.

Fig. 4

Fig. 5

Fig. 6

Fig. 7 Enneper surface

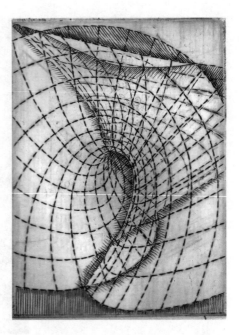

I was interested in a family of surfaces which, since the nineteenth century, bear the name of minimal surfaces. The German school was very active in this field. Weierstraß in 1866 derived a representation formula for this particular family of surfaces:

$$x = \int (1 - \zeta^2)R(\zeta)d\zeta \quad y = \int i(1 + \zeta^2)R(\zeta)d\zeta \quad z = \int 2\zeta R(\zeta)d\zeta \qquad (1)$$

$R(\zeta)$ is a function of a complex variable $\zeta = u + iv$ (i is the square root of -1), which determines the specific minimal surface. Taking the real part of x, y, and z gives the coordinate functions of the surface in the Euclidean 3-space which we also denote with x, y, z.

2.1 Enneper Surface

The first minimal surface I studied with these formulas is that of Enneper (1863) (Fig. 7):

Here $R(\zeta) = 3$, from which one deduces, after integration

$$x = 3\zeta - \zeta^3 \quad y = i(3\zeta + \zeta^3) \quad z = 3\zeta^2.$$

Then after taking the real part we obtain

$$x = 3u + 3uv^2 - u^3, \quad y = -(3v + 3u^2v - v^3), z = 3u^2 - 3v^2.$$

2.2   Formulas of Monge and Weierstraß

Thereafter, I used a generalized formula to represent minimal surfaces ($H^2 = \zeta^2 R$ and $F^2 = R$):

$$x = \int (F^2 - H^2)d(\zeta) \quad y = \int i(F^2 + H^2)d(\zeta) \quad z = \int 2FHd(\zeta) \qquad (1')$$

All these formulas make it possible to compute minimal surfaces starting from an isometric network. The network traced on a surface depends on function "ζ":

- If one takes $\zeta = u + iv$, the conformal representation on the plane (since complex functions preserve angles) maps to a square-like grid on the minimal surface.
- To have a surface bordered by one or more closed curves, one will take $e^\zeta = e^u$ (cos v + i sin v), the conformal representation, here, will be made of concentric circles and radiant lines.

One can also use Monge's formula that one can, for example, deduce from the Weierstraß equations (1) or (1'):

$$dx^2 + dy^2 + dz^2 = 0.$$

This formula is in particular satisfied by the following functions:

$$x = f(\zeta), \ y = g(\zeta), \ z = \int i\sqrt{(f(\zeta)^2 + g(\zeta)^2)}\ d\zeta.$$

The data of the functions F and G implies obviously x and y, and in general with some difficulty z, since computing z requires taking a square root. Since f and g are functions of a complex variable ζ, they define points (x, y) of this plane along a curve. The writing of z, where f and g appear by their square, implies the presence of symmetry at least.

2.3   Catalan Surface

Here is an example established from the cycloid curve in the real plane, which is studied since 1501 (cf: http://www.mathcurve.com/courbes2d/cycloid/cycloid. shtml). The formula of the cycloid curve is:

$$x = u - \sin u, \ y = \cos u$$

After replacement of u by ζ, one obtains the value of z easily:

$$x = \zeta - \sin \zeta$$

Fig. 8 Catalan's minimal
surface

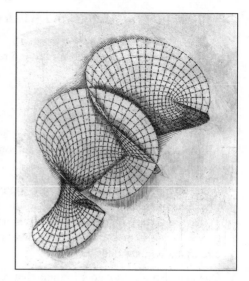

$$y = \cos \zeta$$

$$z = 4i \sin \zeta/2$$

The developed equation gives us:

$$x = u - \sin u \cosh v$$

$$y = \cos u \cosh v$$

$$z = 4 \sin u/2 \sinh v/2$$

The parameters u and v being variable, we obtain the minimal surface of Catalan (1814–1894) (Fig. 8):

2.4 Jeener Surface

There exists many other simply plane curves from which one can obtain minimal surfaces. For example, starting from the planar spiral equation:

$$x = e^{mt} \cos t, \quad y = e^{mt} \sin t$$

one builds the spiral minimal surface (Figs. 9 and 10):

$$x = e^{mt} \cos t$$

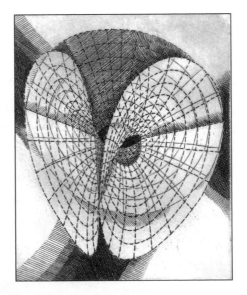

Fig. 9 Minimal surface «à la Chouette»

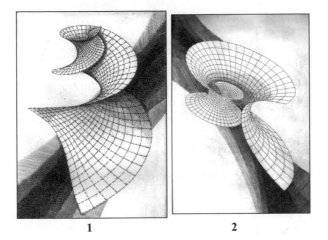

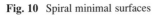

Fig. 10 Spiral minimal surfaces

$$y = e^{mt} \sin t$$

$$z = i \sqrt{((1 + m^2)/m)}$$

If, initially, I chose functions allowing me to trace traditional remarkable surfaces, I then left free course to my imagination. Here is an example:

$$x = u^m \cos mv - (m/(m + 2n)u^{m+2n}) \cos(m + 2n)v$$

Fig. 11 Floraison

$$y = u^m \sin mv + (m/(m + 2n)u^{m+2n})\sin(m + 2n)v$$

$$z = (2mu^{m+n}/(m + n))\cos(m + n)v$$

These surfaces resemble flowers determined by constants m and n (Fig. 11).

2.5 Minimal Surface with a Family of Parabolas

This surface, which comprises a family of parabolas, was studied by Enneper in 1882. It is determined by:

$$R(\zeta) = ia(\zeta^2 - 1)/\zeta^3 - i\, b/2\zeta^2$$

The equation is thus:

$$x = a\, u - a \sin u \, \mathrm{ch}\, v + b \sin u/2 \, \mathrm{sh}\, v/2$$

$$y = a - a \cos u \, \mathrm{ch}\, v + b \cos u/2 \, \mathrm{sh}\, v/2$$

$$z = 4\, a \sin u/2 \, \mathrm{sh}\, v/2 - b\, u/2$$

Fig. 12 Bonnet minimal surface

While making a = 1 and b = 0 one finds Catalan's surface, and for a = 0 and b = 1 the helicoid with planar axis.

2.6 Bonnet Surface

Within the framework of the general research on minimal surfaces having planar principal lines of curvature, Ossian Bonnet discovers a surface (Fig. 12) whose equation is:

$$x = u \cos m + \sin u \text{ ch } v$$

$$y = \sin m \cos u \text{ ch } v$$

$$z = v - \cos m \cos u \text{ ch } v$$

If one takes m = π/2, one finds the catenoid.

2.7 Henneberg Surface

Henneberg discovers the first one-sided resp. unilateral minimal surface (Fig. 13) whose equation is:

$$x = 3 \cos u \text{ sh } v - \cos 3u \text{ sh } 3v$$

$$y = 3 \sin u \text{ sh } v + \sin 3u \text{ sh } 3v$$

$$z = 3 \cos 2u \text{ ch } 2v$$

Fig. 13 Henneberg surface

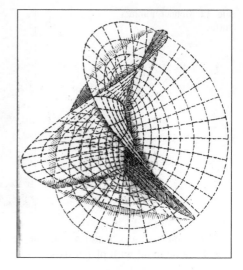

3   Topology of Closed Surfaces Without Singularities

The first closed unilateral surface, the famous Klein bottle, was discovered by Felix
Klein in 1882. The Klein bottle can be generalized to a surface having "n" bottles
given by the following general equation:

$$W = \cos((m + 1)u + \pi/(m + 1)) + 3/2$$

$$x = m \cos u + \cos mu - (m + 1)/2m \; W\sin (m - 1)u/2 \cos v$$

$$y = m \sin u - \sin mu - (m + 1)/2m \; W\cos (m - 1)u/2 \cos v$$

$$z = W \sin v$$

These surfaces are generated by a family of circles whose centres move on
a hypocycloid with n cusps. The surface is unilateral when the number of cusps is
odd: n = 2m + 1 (Figs. 14 and 15).

The simplest generalized bottle is the triple bottle (m = 2, N = 3) (Fig. 16):

$$W = \cos(3u + \pi/4) + 3/2$$

$$x = 2\cos u + \cos 2u - 3/4W \sin u/2 \cos v$$

$$y = 2\sin u - \sin 2u - 3/4W \cos u/2 \cos v$$

$$z = W \sin v$$

Fig. 14 Klein-bottle

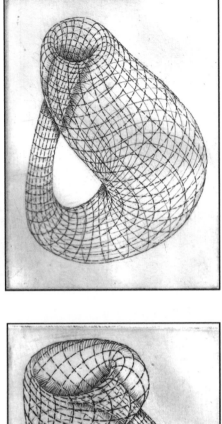

Fig. 15 Double Klein-bottle

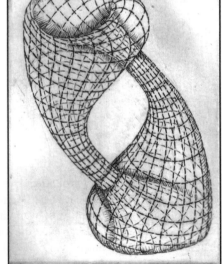

The First Unilateral Closed Surface with only one pole was discovered by Werner Boy in 1902. It has a curve of self-intersection, a three-bladed propeller. There exists a family of these surfaces. They have an odd symmetry. In the case, for

Fig. 16 Triple Klein-Jeener
bottle

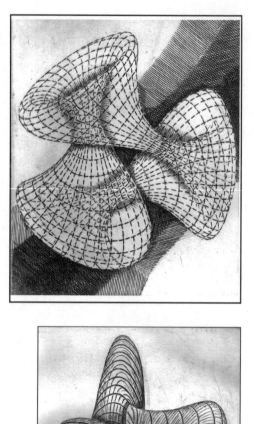

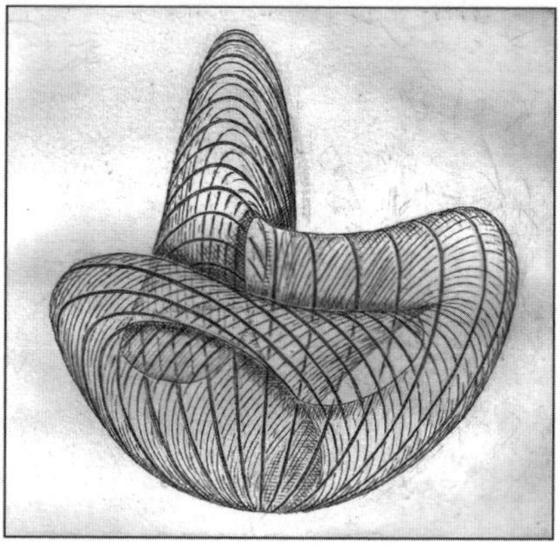

Fig. 17 Boy surface with
symmetry of order 3

example, where symmetry is of order 5, the lines of self-intersection are made up of
two propellers with five blades of different sizes (Figs. 17 and 18).

There is a connected family, of even symmetry. Surfaces of the two families can
be used, in an essential way, to evert the sphere. Bernard Morin introduced and used
that surface whose symmetry is of order 4 and which bears his name. The eversion
occurs at a central stage where two models exist: the open one and the closed. The last
model has a curve of self-intersection, a four-bladed propeller and two circles. Surfaces
of symmetry of order 3 and 4 have a common point: can be traced on each one of
them and entirely recover a set of ellipses. Their equations were published by François
Apéry [1].

These surfaces (Figs. 19 and 20), thanks to their graphic complexity, emphasize
the technique of engraving: indeed, one combines, here, the rigour of a basic

Fig. 18 Boy surface with
symmetry of order 5

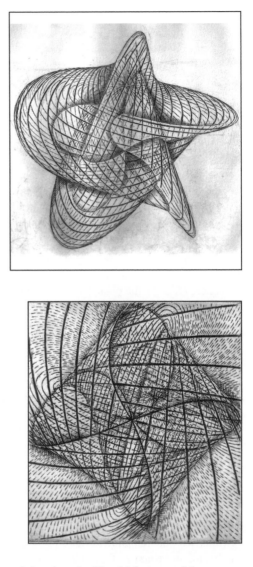

Fig. 19 Apéry model

data-processing layout with that of a work hand made. The thickness and the texture of the curves, and in certain cases, of the lines of self-intersection, must allow a good legibility of surface.

3.1 Surfaces with Constant Total Curvature

Each point on a surface is the intersection of a pair of two orthogonal lines of curvature. The curvature of the surface at this point is the product of the curvatures of the two lines of curvature. When this local curvature is same for each point of the surface then it is said that surface has constant curvature K. Surfaces of this type are

Fig. 20 Morin surface of
order 8

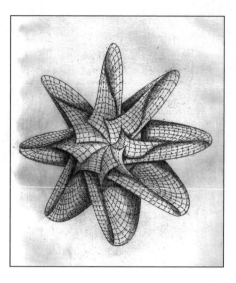

the sphere (K = 1), the plane (K = 0), and the pseudo-sphere (K = −1). Surfaces of negative curvature are called *hyperbolic*.

3.1.1   The Pseudo-sphere (K = −1)

Studied in particular by the Italian mathematician Beltrami in 1868, the pseudo-sphere is a surface of revolution generated by the tractrix, a curve introduced about 1,670 per Claude Perrault (cf. http://www.mathcurve.com/courbes2d/tractrice/tractrice.shtml). Here is the equation of the pseudo-sphere (Fig. 21):

$$x = \cos u \,/ \cosh v$$

$$y = \sin u \,/\cosh v$$

$$z = v - \tanh v$$

Here two other examples where K is equal to a negative constant: Dini surface, a helicoid, and Kuen surface, a 2-soliton.

3.1.2   Dini Helicoid (K = −1)

This helicoid (Fig. 22) is also generated by a tractrix; the second family of curves is formed, here, of circular propellers. The equation can thus take the following form:

$$x = \cos u \,/\cosh v$$

$$y = \sin u/ \cosh v$$

$$z = v - \tanh v + u/4$$

Fig. 21 Pseudo-sphere

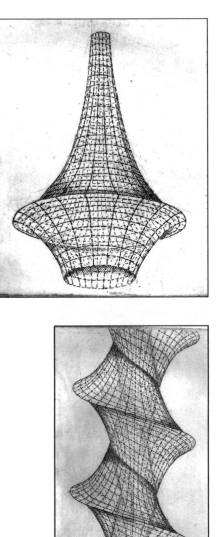

Fig. 22 Dini surface

3.1.3 Kuen Surface (K = −1)

The richness of the surface of Kuen allured several artists. One can see a beautiful model at the Institute Henri Poincaré. It inspired a picture by Luc Bénard [6]. Here is the equation:

$$x = r \cos \theta$$

$$y = r \sin \theta$$

$$z = \log \tan v/2 + a \cos \theta$$

Fig. 23 Kuen surface

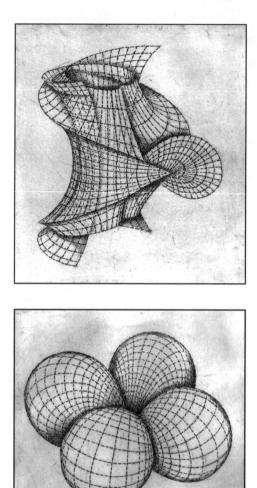

Fig. 24 Sievert's surface

with $\theta = u - \text{arc tan } u$, $a = 2/(1 + u^2 \sin^2 v)$, and $r = a \sqrt{(1 + u^2)} \sin v$ and the engraving I made (Fig. 23).

3.1.4 Sievert Surface (K = +1)

The Sievert surface (Fig. 24) is an example with constant positive curvature $K = 1$. Its equation is:

$$R = 2/\sqrt{(3)}^*(2\sin v \int (1 + 3\sin^2 u))/(4 - 3\sin^2 v \cos^2 u)$$

$$\theta = -u/2 + \arctan(2 + \tan u)$$

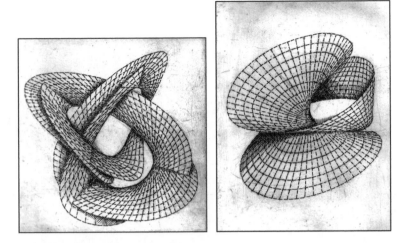

Fig. 25 Skew surfaces

$$x = R \cos \theta$$

$$y = R \sin \theta$$

$$z = 1/\sqrt{(3)}{}^{*}[\log (\tan(t/2)) + 8 \cos t/(4 - 3 \sin^2 t \cos^2 u)]$$

3.1.5 Surfaces of Zero Curvature (K = 0)

Surfaces with $K = 0$ are locally isometric to a plane. Except the cones and the cylinders, these surfaces (Fig. 25) are the loci of the tangents to a given skew curve.

4 Bi-periodical Functions

4.1 Jacobi Functions

Among the complex functions, the first function with two periods was discovered on the basis of the following elliptic integral:

$$F = \int dx/\sqrt{(1 - x^2)(1 - k^2 x^2)}$$

.

Fig. 26 Amplitude function

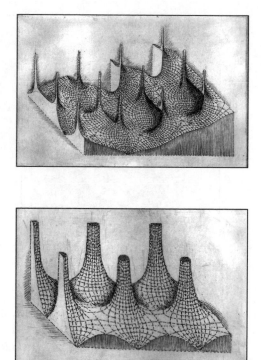

Fig. 27 Sinus-amplitude
function

By inverting this integral, one obtains the Jacobi function called sine amplitude, sn u. There exist two other functions of Jacobi: cosine amplitude, cn u, and delta amplitude, dnu. These functions are connected in the following way: cn u = $\sqrt{(1 - sn^2u)}$ et dn u = $\sqrt{(1 - k^2sn^2u)}$ (Figs. 26 and 27).

4.2 Weierstraß Functions

Weierstraß studies, in its turn, the bi-periodical functions which bear his name. One starts from the integral u = $\int ds/\sqrt{(4s^3 - g_2 s - g_3)}$ where g_2 and g_3 are invariants. e_1, e_2 and e_3 are the roots of the equation $4s^3 - g_2 s - g_3 = 0$.

These functions are denoted by $\wp$ and $\wp'$, where $\wp'$ is the derivative of $\wp$. One thus gets for the first three terms of the series:

$$\wp = 1/u^2 + g_2\ u^2/20 + g_3\ u^4/28$$

$$\wp' = -2/u^3 + g_2u/10 + g_3\ u^3/7 \dots.$$

Fig. 28 Weierstraß
function

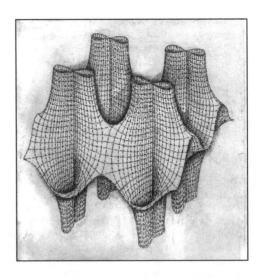

Fig. 29 Weierstraß
function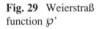

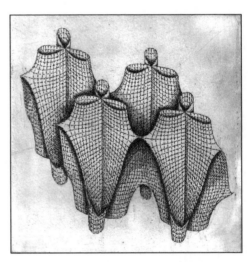

For the engraving of the function $\wp$, one takes $g_2 = 4$ and $g_3 = 0$ (case of the lemniscate) which gives us:

$$e_1 = -1, \; e_2 = 0 \text{ and } e_3 = 1.$$

The two periods are, here, equal, and their value is 2.622... (Fig. 28)

As regard of the function $\wp'$, we choose the equi-anharmonic case $g_2 = 0$ and $g_3 = 1$. Engraving shows well that a section by a plane close to the top cuts the surface according to curves in the clover shape. The plane $z = 0$ cuts the surface according to a tessellation by equilateral triangles (Fig. 29).

Acknowledgement I would like to thank Claude Bruter, Richard Palais, Simon Salamon and in particular Konrad Polthier for their help in the linguistic preparation of the manuscript.
I am currently carrying out the project of engraving an atlas of mathematical models. A list of books [2], [3], [5] which have particularly influenced for my study of surfaces appears in the References. I also took as a starting point the collection of plaster models published by Martin Schilling (Leipzig 1911). One can see, in France, some of these models at the Institute Henri Poincaré and at the Palais de la Découverte.

References

1 Apéry, F.: Models of the Real Projective Plane. Vieweg, Braunschweig-Wiesbaden (1987)
2 Darboux, G.: Théorie Générale des Surfaces. Chelsea, New-York (1986)
3 Fischer, G.: Mathematical Models. Vieweg, Braunschweig-Wiesbaden (1986)
4 Jahnke, E., Emde, F.: Tables of Functions. Dover, New-York (1945)
5 Nitsche, J.C.C.: Lectures of Minimal Surfaces. Cambridge University Press, Cambridge (1989)
6 Palais, R.: Polyhedral Eversions of the Sphere; Gastrulation. In: Bruter, C. (ed.) Mathematics and Modern Art. Springer, Berlin (2012)

Knots and Links As Form-Generating Structures

Dmitri Kozlov

Abstract Practical modeling of spatial surfaces is more convenient by means of transformation of their flat developments made as topologically connected kinetic structures. Any surface in 3D space topologically consists of three types of elements: planar facets (F), linear edges (E) and point vertexes (V). It is possible to identify the first two types of these elements with structural units of two common types of transformable systems: folding structures and kinematic nets respectively.

In the paper a third possible type of flat transformable structures with vertexes as form-generative units is considered. In this case flat developments of surfaces are formed by arranged point sets given by contacting crossing points of some classes of periodic knots and links made of elastic-flexible material, so that their crossing points have real physical contacts. A fragment of plane point surface can be reversibly converted into a fragment of a spatial surface with positive, negative or combined Gaussian curvature by means of transformation which saves connectivity between the points, but not the distances and angles between them. It was proved experimentally that this new form-generative method can be applied to modeling of both oriented and non-oriented differentiable topological 2D manifolds. The method of form-generation based upon the developing properties of periodic structures of knots and links may be applied to many practical fields including art, design and architecture.

D. Kozlov (✉)
Research Institute of Theory and History of Architecture and Town-planning, Russian Academy of the Architecture and Building Sciences, 21-a 7th Parkovaya St., Moscow 105264, Russia
e-mail: kozlov.dmitri@gmail.com

C. Bruter (ed.), *Mathematics and Modern Art*, Springer Proceedings in Mathematics 18, 105
DOI 10.1007/978-3-642-24497-1_10, © Springer-Verlag Berlin Heidelberg 2012

1 Euler's Formula and Two Common Types of Kinetic Surface Models

The most general variety of geometry—topology, treats surfaces in 3D space as 2D manifolds: oriented or non-oriented. It was proved that any oriented manifold is equal to a surface of a pretzel with a some number of holes in it. The number of holes is a topological invariant called "surface genus", which is equal to zero for a sphere, one for a torus, two for a pretzel with two holes and so on. Any 2D surface can be divided into a number of polygonal meshes or facets (F) with borders or edges (E) between them, which intersect in points or vertexes (V). These three elements of an any surface are interrelated by a simple equation known as *Euler's formula*: a number of vertexes minus a number of edges plus a number of facets is equal to two minus two multiplied by n ($V - E + F = 2 - 2n$), there n is the surface genus.

Practical modeling of 2D surfaces in 3D space is more convenient by means of transformation of their flat developments made as connected kinetic structures. There are two well-known types of such structures, which are based upon using planar (F) and linear (E) elements of surface division as their structural invariants. In the first case the result is a folding structure—a flat solid sheet divided into a number of planar facets (F) with turning linear (E) hinges between them (Fig. 1). Flat kinetic folding structures are the basic form-generative principle for different types of transformable paper models, including the art of origami.

In the second case the structure is a kinematic net—a flat net with non-triangle meshes assembled of linear (E) elements with turning point hinges (V) between them. The net can completely lie on a plane or be transformed fully or partly into spatial position (Fig. 2). The principle of kinematic net structure has found its wide application in practical modeling of complex curved surfaces. In 1878 Russian mathematician P. L. Chebyshev stated equations for flat developments of spherical surfaces made of fabric with square meshes [1]. In the end of nineteenth century A. Gaudi used the method of inversion of suspended net models with the aim of

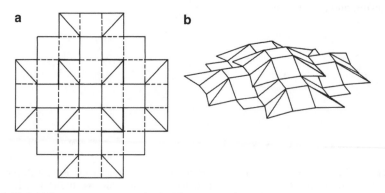

Fig. 1 Flat folding structure as a method of form-generation of surfaces

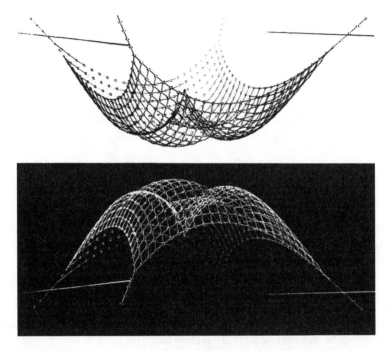

Fig. 2 Kinematic net with square meshes as a method of form-generation of surfaces

form-finding in architecture. In the middle of twentieth century F. Otto started his own experiments with suspended net models that lead him and his colleagues to a new approach to grid shells building theory and a whole number of architectural masterpieces [2].

2 Vertex Structures as a New Third Type of Kinetic Surface Models

In addition to the planar and linear types of flat developments of surfaces it may be proposed a third possible type of flat transformable structures with vertices (V) as form-generative units. Approximations of a surface by number of points is a common method in mathematics and computer graphics. A separate point in this case is just a dot in virtual space determined by its numerical value in relation to three Cartesian coordinates.

A physical model of a point can be done as a contact of two physical bodies such as tangent solid spheres or tangent cylinders with non-parallel axis. A number of contact points on a plane or in space may be represented as a vertex or point surface (Fig. 3), but to function as a transformable model of continual surface the contacting bodies must be connected between them and organized into a kinematic

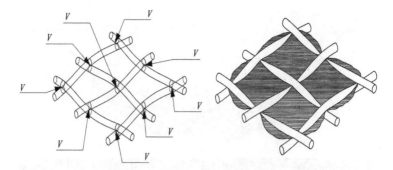

Fig. 3 A fragment of point surface made of woven resilient rods

structure. The structure is the most important part of point models of surfaces because it coordinates behavior of great number of contact points to provide them with the possibility of synchronized sliding movement.

This structure is not just a simple sum of neighboring kinematic units like the structures of planar (*F*) and linear (*E*) models of surfaces—it is *synergetic* in the R. B. Fuller's meaning of the word: a "behavior of integral, aggregate, whole systems unpredicted by behaviors of any of their components or subassemblies of their components taken separately from the whole" [3, p. 3].

3   Resilient Knots and Links as a Structural Principle of Vertex Surface Models

My own experimental research into different plain vertex models confirmed that the most natural forms of organizing independent point contacts into topologically connected structures are knots and links [4]. A resilient rod forms an elementary structure then its ends are joined together. As a result the rod becomes a ring—a trivial knot (Fig. 4a), and its structural stability depends on the ratio between the diameter of the ring and diameter of cross-section of the rod. Then the diameter of the ring is too large to resist the inner torsion forces in the bent rod, the ring turns into double nested loops (Fig. 4b). If the process of loops emerging is combined with joining together of the free ends of the rod, the connected rod may be knotted and take form of the simplest knot "trefoil" (Fig. 4c).

The process of "self-knotting" is very typical for long flexible-resilient strings, such as steel wire or fishing-line. Natural string-like flexible long objects, such as polymeric molecules including DNA, often take circular closed forms of rings and knots either single or linked [5]. Knots and links are widespread and natural way of structural organization for string-like flexible-resilient long objects.

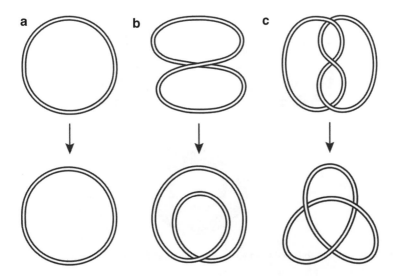

Fig. 4 (a) Resilient ring (trivial knot). (b) Double loop (trivial knot). (c) Simplest knot—trefoil

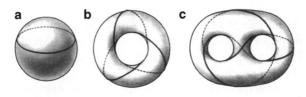

Fig. 5 Knots on 2D manifolds

4 Knots on Different 2D Surfaces

The trefoil is a "torus knot", because it can be placed without any self-crossings on the surface of a torus (Fig. 5b). Like a trefoil, there are knots that can be placed on the surfaces of other 2D manifolds: a ring or trivial knot on a sphere (Fig. 5a), "figure eight" knot—on the surfaces of pretzels with two holes (Fig. 5c) and so on.

The trefoil knot may have two mirror types—a "left" one and a "right" one. Each of them can be tied on the torus surface without self-crossings (Fig. 6a, b), but been tied together on the same torus, they inevitably have contact points between them and form a knotted fabric on torus surface (Fig. 6c). If both knots made of resilient material and their crossings are really contacting, the structure will represent a model of torus point surface.

The contacting points define the model of the surface—namely the exterior shape, and two mirror knots form its interior structure. In the same way it is possible to receive a point surface of an arbitrary pretzel with two mirror pretzel knots of appropriate type.

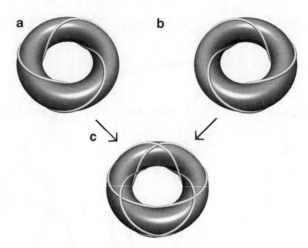

Fig. 6 Two mirror trefoils tied together on the same torus form a torus point surface

5 Energy of Resilience as Forming Principle of Cyclic Knots

Quantity of elastic energy or energy of resilience in a knotted rod depends of topological complexity of a knot and is known among other topological invariants of knots [6]. Thanks to this energy the central lines of knotted rods tend to coincide with a plain, so all their crossings tend to be really contacted, that let them form a model of flat point surface. The two mirror trefoils on the torus surface also tend to collapse, and if the torus itself disappeared, the contacting points of the two knots would place themselves in a flat ring-shaped area. And vice versa: a flat model of point surface, given by a torus knot or a link, may be transformed into a spatial state and fixed in it in order to keep the received shape.

The energy of resilience in knots also defines geometry of their structures. It force a closed resilient rod to take a shape of a ring and a rod of the same material knotted into a trefoil—a shape of a double turn coil. A coil is a natural shape for any knotted and closed resilient rod defined by its minimal internal energy of resilience. At the same time for some periodic knots [7] like trefoil, their coils may be divided into a number of equal loops or "petals": in the case of trefoil the number of loops is three. Knots of this type with natural numbers of coil turns and petal loops have a general name of *Turk's Heads*. Simple Turk's Heads with small numbers of coil turns and petal loops made of soft non-resilient material such as rope, have a wide spread in seamen's practice as well as in the field of decoration and art [8]. It is possible to give the name of *cyclic knots* to the Turk's Head knots made of resilient material because of geometrical structure of their shape. From the topological point of view they are periodic closed braids [9].

All cyclic knots and links can be classified according to the numbers of their turns of coils (p) and petal loops (q) in the system of orthogonal coordinates. The

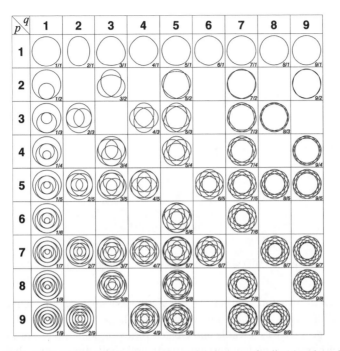

Fig. 7 Cyclic knots classified according to numbers of their turns of coils (p) and petal loops (q)

numbers p and q may be equal to any natural number: if they are coprime—the structure is a knot (Fig. 7). Here is exactly the same law as for epicycloids: diameters of their generating circles must be coprime natural numbers. If p and q are not coprime numbers the structure is a link of equivalent knots, and number of linked knots is equal to the greatest common divisor of p and q.

6 Form Generative Properties of Cyclic Knots and Links

The possibility to transform a cyclic knot from flat position to a spatial one depends of a sufficient number of its contacting crossings. This number is determined by numbers of turns of coils and petal loops, and consequently of total resilient energy of knotted rods. Increasing of the energy proportionally to the quantity of contacting crossings leads knots to a new property: from simplest knots like a trefoil they grow into complicated structures that can serve as models of point surfaces. I gave the name *NODUS* structures to these cyclic knots designed specially for modeling of planar and spatial point surfaces (the word *nodus* means *a knot* in Latin) [10].

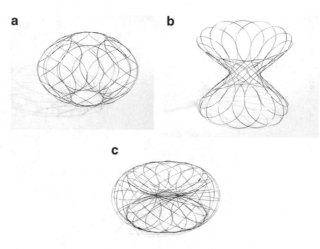

Fig. 8 NODUS structures with elliptic, hyperbolic and combined surface curvatures

A NODUS structure during its transformation changes the lengths of the edges of all its facets and angles between them. Thanks to that ability, the structure changes its geometry as a whole and creates vertex or point models of the surfaces with an arbitrary Gaussian curvature: parabolic, elliptic or hyperbolic. These three types of surfaces completely exhaust all possible internal geometries of two-dimensional manifolds [11]. As contrasted to solid models of surfaces, that can not change their Gaussian curvatures without breaks and folds, point surfaces of NODUS structures permit transition from positive Gaussian curvature (elliptic) to negative one (hyperbolic) through mediation of neutral (parabolic) curvature. The same NODUS structure can take forms of elliptic and hyperbolic curvature. A surface of torus is a combination of these two types of curvatures together with two intermediate areas of parabolic curvature (Fig. 8a–c).

A surface of pretzel may be received as a combination of several torus structures (Fig. 9a). It is possible to create many other forms, for example surfaces with self crossings (Fig. 9b,c).

Also NODUS structures let make fragments of non-oriented 2D manifolds such as Möebius band with self crossing (Fig. 10) or cross-cap—part of projective plane [12].

Apart from the transformation of NODUS structures that changes the sign of its curvature and which can be named *qualitative transformation*, there is another kind of transformation—the *quantitative* one. This transformation happens as a gradual changing of numerical value of Gaussian curvature of point surface from its minimum to a maximum value without an alteration of the curvature sign. The minimum value of Gaussian curvature may be equal to zero, and in this case the point surface of a NODUS structure approximates a piece of plane. In this case the process of transformation represents a continual sequence of changing forms, for example from spherical segment through hemisphere to sphere (Fig. 11a–d). The transformation of NODUS structure is a reversible process.

Fig. 9 Different forms of surfaces received by means of NODUS structures

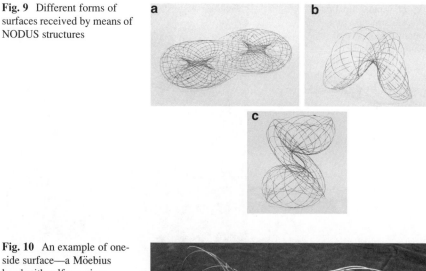

Fig. 10 An example of one-side surface—a Möebius band with self crossing

Thanks to its form changing, NODUS structure accumulates elastic energy and becomes stronger. Every spatial form of NODUS structure may be strictly fixed by limitation of its mobility and as a result the transformable structure will become a stable one.

Polymorphous properties of NODUS structures give to an artist or a designer a suitable tool not only for finding the demanded form in space, but also for "tuning" it in environment. It is possible to envision in advance a script of development of a planar point structure into a surface in three-dimensional space by means of different dispositions of modular form-generating structures on a plane, by choice of their connections and by spatial stratifications of their contact points.

According to my experiments, NODUS structures allows to extrapolate their structural properties from models to large-sized structures (Fig. 12), that gives a reason to consider them also as form-generative principle for real-size kinetic architectural structures [13].

Fig. 11 Transformation of
NODUS structure as
continual sequence of
changing shapes

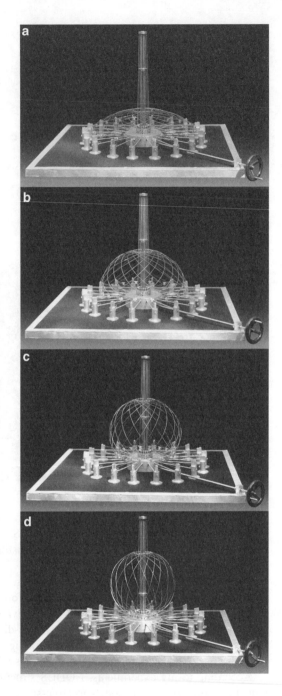

Fig. 12 A large-sized
NODUS structure in natural
environment

References

1. Chebyshev, P.L.: Sur la Coupe des Vêtements. In: Chebyshev, P.L. (ed.) Complete Works (in Russian), vol. V, pp. 165–170. Academy of Sciences of USSR, Moscow-Leningrad (1951)
2. Otto, F., et al.: Grid Shells (IL-10). Institute for Lightweight Structures, University of Stuttgart, Stuttgart (1974)
3. Fuller, R.B.: Synergetics. Macmillan, New York (1982)
4. Kozlov, D.: Synergetic Structures of Topological Knots and Links as Physical Models of Point Surfaces in 3D Space, p. 52, 1036th AMS Meeting. New York University, New York (2008)
5. Liu, L.F., Depew, R.E., Wang, J.C.: Knotted single-stranded DNA rings. J. Mol. Biol. **106**, 439–452 (1976)
6. Stewart, I.: Finding the Energy to Solve a Knotty Problem. In New Scientist, p. 18, March (1993)
7. Burde, G., Zieschang H.: Knots. Walter de Grujter, Berlin (1985)
8. Ashley, C.W.: The Ashley Book of Knots. Faber & Faber, London (1993)
9. Manturov, V.: Knot Theory. Chapman & Hall/CRC, Boca Raton (2004)
10. Kozlov, D.Yu.: Polymorphous resilient-flexible shaping structures "NODUS" for space and other extreme environments. In: Final Conference Proceedings Report of The First International Design for Extreme Environments Assembly, pp. 259–260. University of Houston, Houston (1991)
11. Hilbert, D., Cohn-Vossen, S.: Geometry and the Imagination. Chelsea, New York (1999)
12. Francis, G.K.: A Topological Picturebook. Springer, New York (1987)
13. Kozlov, D.Yu.: Dome structures for flexible material. In: Roofs: Human Settlements and Socio-Cultural Environment, Part 1, UNESCO, vol. 3, pp. 127–131, Paris (1991)

Geometry and Art from the Cordovan Proportion

Antonia Redondo Buitrago and Encarnación Reyes Iglesias

Abstract The Cordovan proportion, $c = (2-\sqrt{2})^{-1/2}$, is the ratio between the radius of the regular octagon and its side length. This proportion was introduced by R. de la Hoz in 1973. Recently, the authors have found geometric properties linked with that proportion, related with a family of shapes named by them, *Cordovan polygons*. These results are summarized and are extended through the works of art of Hashim Cabrera and Luis Calvo, two Cordovan painters who have consciously considered the Cordovan proportion in their recent compositions. In fact, we have checked this ratio in several dissections of the canvases of Cabrera, and looking at the picture of Calvo, we have recognized many of our Cordovan polygons and some new polygons which we have added to our previous collection. We have also discovered some new cordovan dissections of a square, a $\sqrt{2}$ rectangle and a Silver rectangle.

1 Introduction

This work is the result of a meeting of two mathematicians with two painters in Córdoba city in Spain, and its conversations about the Cordovan proportion.

The Cordovan proportion, $c = (2-\sqrt{2})^{-1/2}$, is the ratio between the radius, R, of the regular octagon and its side length, L (Fig. 1).

A.R. Buitrago (✉)
Departamento de Matemáticas, I.E.S. Bachiller Sabuco, Avenida de España, 9 – 02002 Albacete, Spain
e-mail: aredondo@sabuco.com

E.R. Iglesias
Departamento de Matemática Aplicada, Universidad de Valladolid, Avenida Salamanca, s/n – 47014 Valladolid, Spain
e-mail: ereyes@maf.uva.es

C. Bruter (ed.), *Mathematics and Modern Art*, Springer Proceedings in Mathematics 18, DOI 10.1007/978-3-642-24497-1_11, © Springer-Verlag Berlin Heidelberg 2012

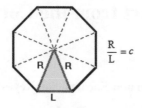

Fig. 1 Cordovan proportion in the regular octagon

This proportion was introduced by the Spanish architect Rafael de la Hoz Arderius in 1973 [1, 2] and named by him, *Cordovan proportion*. The irrational value of this ratio is known as the *Cordovan number*. By means of the law of Cosines, in the marked triangle with sides R, R and L (see Fig. 1), we have:

$$L^2 = 2R^2 - 2R^2 \cos 45° = R^2\left(2 - \sqrt{2}\right) \Rightarrow \frac{R^2}{L^2} = \frac{1}{2 - \sqrt{2}}$$

$$c = \frac{R}{L} = \frac{1}{\sqrt{2 - \sqrt{2}}} = 1.306562964... = Cordovan\ Number$$

Before 2008, when the authors began the study of this proportion, only the rectangular shape had been considered. From that year, their research has found geometric properties linked with that proportion, related to a wide family of new forms named by them, *Cordovan polygons* [3–5].

In this contribution, those results are summarized and are extended through the works of art of two Cordovan painters which have consciously considered the Cordovan proportion in their recent compositions. The aims, motivations and visual results are completely different within their works.

2 Polygonal Shapes and Cordovan Proportion

In this section, we summarize the main polygonal shapes discovered and related with the Cordovan proportion.

The *Cordovan triangle* is the isosceles triangle which is similar to the one in Fig. 1 of sides R, R and L. So, an isosceles triangle is a "Cordovan triangle" if its angles are $\pi/4$, $3\pi/8$ and $3\pi/8$ radians (Fig. 2).

The *Cordovan rectangle is a rectangle whose sides are in ratio c*. For instance, the marked rectangle of sides R and L showed in Fig. 3.

The *Cordovan diamond is a rhombus whose angles are* $\pi/4$, *and* $3\pi/4$ radians. This shape is formed by the union of two Cordovan triangles. Four octagons intersected as in Fig. 4 to produce an inner star formed by four diamonds.

The regular octagon can be divided into four congruent quadrilaterals, named by the authors *Cordovan kites or c-kites*. A c-kite is formed by two Cordovan triangles.

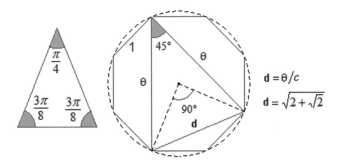

Fig. 2 Cordovan triangle and its location within the regular octagon

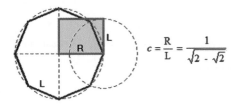

Fig. 3 Cordovan rectangle and its construction from a regular octagon

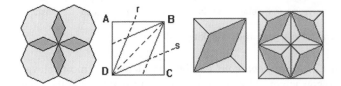

Fig. 4 Diamonds in octagons and diamonds in a square by bisecting 90° and 45°

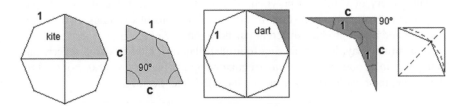

Fig. 5 Cordovan kite, Cordovan dart and its construction from the square

This quadrilateral has angles $\pi/2$, $3\pi/8$, $3\pi/4$, and $3\pi/8$. If the side of the octagon is 1, the c-kite has sides 1, 1, c, and c, Fig. 5.

When the octagon is inscribed in a square, another four congruent concave quadrilaterals (see Fig. 5) appear. These quadrilaterals are called as *Cordovan darts or c-darts*. A c-dart has angles $\pi/8$, $\pi/2$, $\pi/8$ and $5\pi/4$, and sides 1, 1, c, and c.

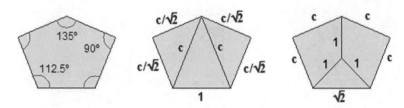

Fig. 6 A Cordovan pentagon

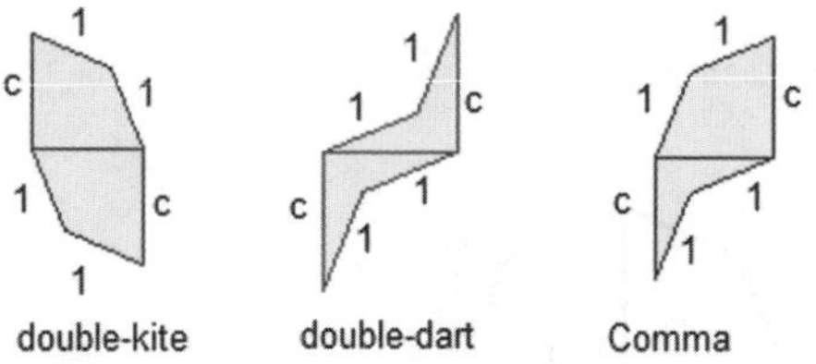

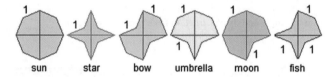

Fig. 7 Cordovan par-hexagons

double-kite   double-dart   Comma

sun   star   bow   umbrella   moon   fish

Fig. 8 Several par-octagons made with kites and darts

Several *Cordovan pentagons* can be considered, the most relevant is the non regular polygon achieved by means of two right triangles over the equal sides of the Cordovan triangle (Fig. 6). This pentagon covers the plane. It has four equal sides of $c/\sqrt{2}$ and the remaining side is 1. Its angles are $5\pi/8$, $\pi/2$, $3\pi/4$, $\pi/2$ and $5\pi/8$.

The gallery of Cordovan polygons is wide [5]. Indeed, the collection of par-polygons is especially interesting. In fact, combinations of two kites, two darts or one kite and a dart produce different par-hexagons (see Fig. 7).

In a similar way, by considering four quadrilaterals, kites or darts, several par-octagons can be obtained: the *c-sun* (regular octagon), the four point *star* or (*c-star*), the *c-bow*, the *c-umbrella*, the *c-moon*, the *c-fish*, etc. See Fig. 8.

3 Hashim Cabrera: The Proportion of the Soul

Hashim Cabrera was born in Sevilla in 1954, soon afterwards he moved to Almodovar del Rio (Córdoba, Spain) where he now lives. He is a painter and writer interested in both visual arts and its historic, spiritual and philosophical basis.

After a first stage, where the artist was concerned about the form, he evolved towards a *naturalist abstract* concept of the painting. By avoiding shapes and symbols, the artist's message can be freed of its servitude to the form. Under these assumptions, *colour* and *proportion* are the only elements allowed.

In Córdoba, in February 2008, he exhibits *Los colores del alma* (*Colours of the soul*), ten paintings (acrylic/canvases) which are a sample of his profound reflection about colour. All works are conceived as the union of two, three or four rectangular modules; each of them in a single colour. The complete exposition can be found in http://www.hashimcabrera.com/galeria.html.

Cabrera claims *since we can't live an absolute oneness permanently, then oneness and diversity can't be separate*. This idea is represented by the main use of green, the only primary colour which at the same time is secondary. The harmony is achieved by means of a balance between the pieces. The manner of the division of the support space is purposely arranged by the Cordovan proportion. In addition, all compositions are also interrelated by means the Cordovan proportion.

Surprisingly, the Cordovan number appears to be involved in the measured values for the annual mean sunlight in Cordoba. This striking fact attracts attention of the painter, who is fascinated by that coincidence which joins an external phenomenon, macrocosmic, with an inner experience based in the mathematical concepts which arrange the order and consensus. That is, the Cordovan proportion.

We have checked and we have found this ratio in the dissections of the canvases the Cabrera's exposition, even in those cases where it is not evident. As a sample, we will focus in the paintings showed in Fig. 9. From these, some geometric facts are found.

The picture in Fig. 9a is an almost perfect square, formed by the union of two modules which shape the simplest Cordovan dissection of a square:

A square may be divided into a Cordovan rectangle and a rectangle of ratio c/(c-1)

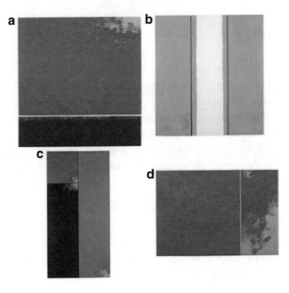

Fig. 9 Four canvases of Hashim Cabrera (Photographs by Bruno Rascado)

Fig. 10 Cordovan division on *Diptych chrome green /black*

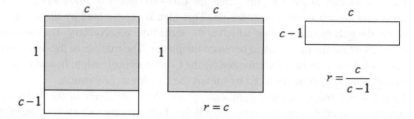

Fig. 11 A square is the union of a Cordovan rectangle and a rectangle $c/(c-1)$

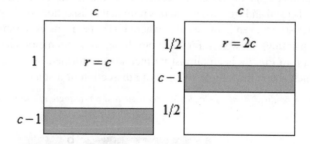

Fig. 12 Dividing the square into rectangles with ratio $2c$ and $c/(c-1)$

The graphic analysis of the canvas is showed in Fig. 10. On the left, we can see the painting. In the centre we observe the aforementioned dissection of the square, made with accuracy by means of a geometric design program. On the right, through a simple superposition of the preceding images it is shown the "almost perfect" Cordovan dissection achieved by Cabrera.

Starting with the dissection of Fig. 11, we can move the rectangle to place it in the centre of the square. The result is another Cordovan dissection of the square, Fig. 12:

A square can be divided into one rectangle of ratio $c/(c-1)$ and two rectangles with ratio $2c$

We have recognized this dissection in "Visitors", where we observe a white central strip dividing symmetrically the canvas. Precisely, that strip is a rectangle with ratio $c/(c-1)$.

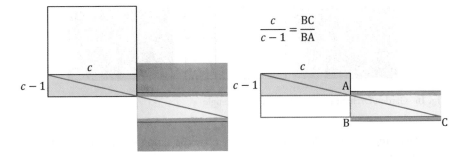

$$\frac{c}{c-1} = \frac{BC}{BA}$$

Fig. 13 Cordovan division on *Visitors*

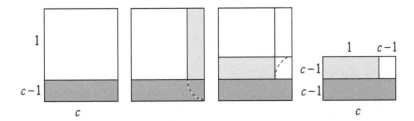

Fig. 14 Construction and dissection of the rectangle $c/(2c\text{-}2)$

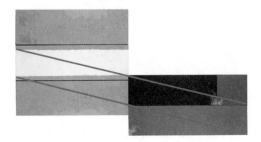

Fig. 15 A rectangle $c/(2c\text{-}2)$ in *Triptych green/black*

Figure 13 explains without words the graphic analysis of this painting showed in Fig. 9b.

Starting again by the first division of the square, we can achieve an harmonic dissection of a rectangle of ratio $c/(2c\text{--}2)$, Fig. 14.

This pattern appears in Triptych green/black, Fig. 9c. This assertion is explained in the Fig. 15. Observe that both blue lines are parallel and the upper line is the extension of the diagonal of the white rectangle $c/(c\text{--}1)$.

Figure 9d shows *Fruit garden*, this composition is surprising. The painting is formed by a squared module adjacent to another rectangular one. It is the only case where the artist recognized that the proportions in this composition are considered by chance. It is striking that even in this case the Cordovan proportion turns up. See Fig. 16.

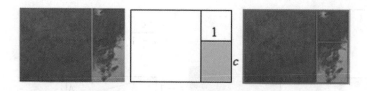

Fig. 16 Garden/Orchard towards the Golden rectangle

Indeed, the rectangular module is formed by the union of one square and one Cordovan rectangle.

We discover the two first steps of a pseudo-gnomonical growth associated to a Fibonacci sequence $G_n = G_{n-2} + G_{n-1}, n = 2, 3, 4, \ldots$ with initial terms $G_0 = c$ and $G_1 = 1$:

$$c, 1, c + 1, c + 2, 2c + 3, 3c + 5, \ldots$$

Observe that

$$\frac{G_{n+1}}{G_n} \to \phi \ (Golden \ mean) \text{ and } G_n = F_{n-1}c + F_n, \ n = 1, 2, 3, \ldots$$

where F_n is the term of the usual Fibonacci sequence $F_n = F_{n-2} + F_{n-1}, n = 2, 3, 4, \ldots$ which starts in $F_0 = 0$ and $F_1 = 1$. So the intuitive shape by the painter may be considered as an approximation of the Golden rectangle.

4   Luis Calvo: A Vectorial Drawing of the Mosque of Córdoba

Luis Calvo born in 1959 is a designer, an ardent admirer of the history and architecture of Córdoba, his home town. In December 2008, the Cervantes Institute in Brussels organized an art exhibition devoted to intra-cultural expressions of Córdoba city, through the work of eight artists. Calvo contributed to the collective exhibition with the painting, "La Proporción Cordobesa" (The Cordovan proportion), Fig. 17, where the author gives homage to the Mosque, the most emblematic building of the city. Calvo's work is a vectorial drawing on cotton fabric, on frames which are arranged forming a rectangle, which represent the plan of the Mosque of Cordova.

Looking at the picture, we can see five rectangular modules, each of them coloured, differently in accordance to the different stages of the construction of the building. On the full rectangle, we observe many octagonal stars. Each one is inserted in a geometric framework formed by a rectangular grid jointly with another grid which has been rotated with respect to the first one.

Let us focus our attention on the geometric skeleton where the octagonal stars are placed. Its mathematical interest becomes evident when we construct it.

Fig. 17 La Proporción Cordobesa (2008)

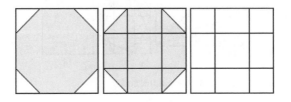

Fig. 18 Basis of the initial grid

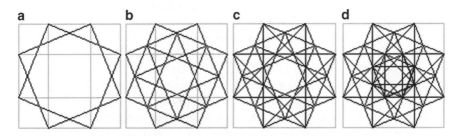

Fig. 19 Generating the basis star

Dissecting a regular octagon inscribed in a square, using lines that are parallel to the sides of the square, Fig. 18, we may determine a compound figure formed by four congruent rectangles of ratio $\sqrt{2}$, four congruent squares and a big central square. This module will be the basis of the initial grid.

The first figure to be constructed is the star octagon, denoted by 8/2, (Fig. 19a), which is formed by two squares. In the next step, inside the previous star, we draw the star octagon 8/3, or octagram, with its points lying in the convex vertices of the star 8/2, (Fig. 19b).

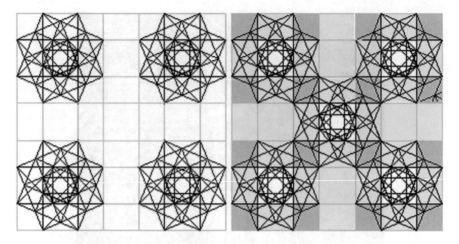

Fig. 20 Generating the pattern basis

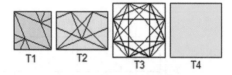

Fig. 21 Rectangular and squared modules in the grid

Returning to the first star 8/2, we draw another star 8/3 with its points in the points of the initial star 8/2. This operation determines eight interlaced Cordovan triangles with its angles of 45° in the corners of the stars (Fig. 19c). Within the little internal star 8/2 determined in the centre of the composition, we repeat the same operations described in the previous lines. This composition is translated over the grid in two independent directions, Fig. 20 (left). Finally, we install another identical design over the inner canonical module surrounded by the four previous modules. The final result is a grid of five interlaced modules, Fig. 20 (right).

By means of two independent vectors, we can translate the five compositions in order to achieve the picture of a tiling of rectangles and squares, some of them decorated by segments involved in the octagon geometry, Fig. 21.

It is important to remark that the artist has used a very different construction procedure. The point of start of Calvo is the strategic placement of the T1 module, in the small squares of a rectangular grid as in Fig. 18. Next, the segments are extended and the drawing is completed by adding the remaining lines. See Fig. 22.

In the T1 module, the Cordovan diamond appears purely inscribed in a square constructed by bisecting the angles of 90° and 45° (see Fig. 4). Let us observe that this shape is one of the two bases of origami: *the Fish base*. In the eighth century, the art of paper folding was very popular among the Arab people. The personal proposal of Calvo is that the Mosque of Córdoba was constructed by master

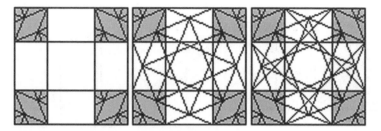

Fig. 22 Construction from T1

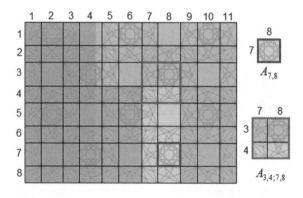

Fig. 23 Codifying the canvas

bricklayers who knew and used origami techniques. This is the reason for the main role of the module T1 in the way followed by the artist.

Through both preceding alternative constructions, the final result is the same, a modern "tastir" (straight line geometry). The Cordovan proportion emerges from this gallery of stars, where several Cordovan polygons are easy recognizable, which have not been drawn consciously, but appear in an implicit way. We discovered in the grid some new polygons which we could be added to our previous collection.

In order to clarify the understanding of the canvas, we are going to consider it as an array of eight rows and eleven columns, and we will codify each one of the boxes or set of boxes, as explained in Fig. 23. The background of this figure is a previous sketch of the painting kindly provided by the author.

The Cordovan diamond is found off in several boxes. For example in $A_{i,j}$, with $I = 2, 4, 6, 8$ and $j = 1, 3, 5, 7, 9, 11$. The Cordovan triangle is spread around the picture. For instance, in the box $A_{4,5;2,3}$, we recognize two Cordovan triangles, respectively divided by each other and its respective gnomon. A *c-bow* is located in $A_{3,4;3,4}$. A concave octagon, known as *star* and a *c-umbrella* are located explicitly in $A_{3,4,5;2,3,4}$. There are octagons *c-fish* and *c-moon* in $A_{5,6,7;4,5,6}$, and so on, Fig. 24.

By analyzing the picture, we can define about some new pentagons which will complete the family of pentagons discovered by the authors. These new Cordovan pentagons are cut into rectangles, squares, cordovan triangles and gnomons, see Fig. 25. The first pentagons involves the Cordovan number and the Silver number

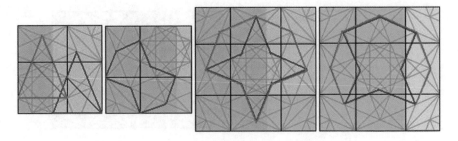

Fig. 24 Cordovan polygons in boxes $A_{4,5;2,3}$, $A_{3,4;3,4}$, $A_{3,4,5;2,3,4}$ and $A_{5,6,7;4,5,6}$

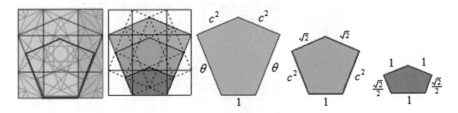

Fig. 25 Three pentagons in the box $A_{7,8,9;6,7,8}$

Fig. 26 Dissections of the rectangles $\sqrt{2}$

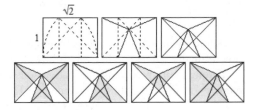

$\theta = 1+\sqrt{2}$. They appears in the rows one and two passing for several columns: $A_{1,2;1,2,3}$, $A_{1,2;5,6,7}$, $A_{1,2;9,10,11}$. Symmetric pentagons of these ones can be found in $A_{4,5;1,2,3}$, $A_{4,5;5,6,7}$,, $A_{4,5;9,10,11}$.

Looking at the T2 module, we discover several geometric facts about the rectangle $\sqrt{2}$. The upper row of Fig. 26 explains the steps of construction of T2 module. In the first picture of the procedure becomes evident that *a rectangle $\sqrt{2}$ can be divided into a Silver rectangle and two rectangles of ratio c^2.*

In the bottom row, on the left, we can see another dissection of the rectangle $\sqrt{2}$. In this case the result is the union of three Cordovan triangles, an isosceles triangle and two symmetric scalene triangles. That isosceles triangle is the gnomon of the scalene triangle, which is precisely the gnomon of the Cordovan triangle. The remaining pictures show the location of a c-dart and a c-kite in the rectangle $\sqrt{2}$.

5 Conclusions

We can conclude that the Cordovan Proportion is not just a mathematical invention. It is not a simple matter to coin a new geometric term, but it is considered by artists like Cabrera and Calvo in their art works. Research in two very different fields, mathematicians and artists converge to a common point: The Cordovan Proportion, the ratio which emerges from the soul of the Mosque of Córdoba.

Acknowledgements The authors express their hugely grateful to the Cordovan artists, Hashim Cabrera and Luis Calvo, who have given their permission to the authors of this paper to use their art works. Indeed, this paper would not have been written without the existence of their paintings. Thanks also for allowing us to use the photographs of Bruno Rascado.

References

1. Hoz, R.: La proporción Cordobesa. Actas de la quinta asamblea de instituciones de Cultura de las Diputaciones. Ed. Diputación de Córdoba (1973)
2. Hoz, R.: Rafael de la Hoz. Consejo Superior de los Colegios de Arquitectos de España. ISBN 8460977234. Córdoba (Spain) (2005)
3. Redondo, A., Reyes, E.: The cordovan proportion: geometry, art and paper folding. Proceedings of 7th Interdisciplinary Conference ISAMA 2008-Valencia, pp. 107–114, Spain, Hyperseeing, May–June (2008a)
4. Redondo, A., Reyes, E.: The geometry of the cordovan polygons. Vis. Math. **10**(4) (2008b), www.mi.sanu.ac.yu/vismath/redondo2009/cordovan.pdf
5. Redondo, A., Reyes, E.: Cordovan geometrical patterns and designs. Symmetry: Art and Science, 009/1–4. Special Issue for the Conference of ISIS-Symmetry. Symmetry of Forms and Structures, pp. 68–71, Wroclaw-Cracow, Poland (2009)

Web Pages

http://www.hashimcabrera.com/galeria.html
http://www.lacajadelagua.com/la_sexta_mirada/aut_luis.html

Dynamic Surfaces

Simon Salamon

Abstract After discussing some well-known examples in geometry and function theory, we study surfaces in space that are defined by the vanishing of the torsion of integral curves of a given vector field.

1 Introduction

This is a record of a lecture given at the ESMA conference on mathematics and art at the Institut Henri Poincaré in Paris on 20 July 2010. The aim was to present a number of striking images and animations based on the application of techniques from both differential geometry and dynamical systems. This article retains the title of the lecture, even though it can only present still images; the true dynamic content can be found from links in the references at the end of the article.

A typical example is the opaque surface visible in Fig. 1 "supporting" the celebrated Lorenz attractor. The surface consists of points in space at which the corresponding trajectory has zero torsion, a concept that will be intuitively described in Sect. 3 and that is defined in any elementary course on the differential geometry of space curves. Yet the surface is defined by an extremely complicated equation of degree 8 (reproduced as Fig. 12), whose significance could not be recognized without the underlying theory. Like the study of fractals, this is primarily based on scientific discovery, but the cataloguing of the resulting images has an artistic aspect to it. The point of this article is to give a glimpse of both the theory and the imagery to a wide audience.

To introduce the subject of differential geometry that underlies the approach, Fig. 2 is the author's photograph of part of a lecture by Richard Hamilton in Pisa in 2004. Whilst the blackboard relayed advanced mathematics to an expert audience

S. Salamon (✉)
Politecnico di Torino, Corso Duca degli Abruzzi 24, 10129 Torino, Italia
e-mail: salamon@calvino.polito.it

C. Bruter (ed.), *Mathematics and Modern Art*, Springer Proceedings in Mathematics 18,   131
DOI 10.1007/978-3-642-24497-1_12, © Springer-Verlag Berlin Heidelberg 2012

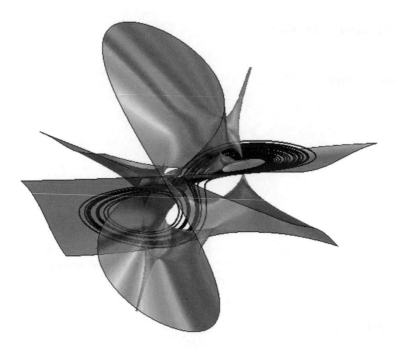

Fig. 1 Lorenz surface

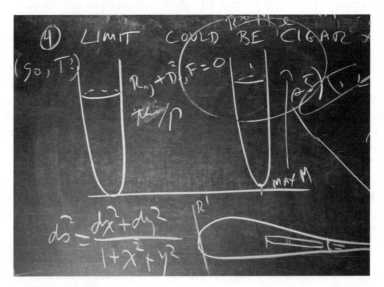

Fig. 2 Discussion of Ricci flow

of conference participants, it conveys even now the essence of mathematical communication carried out part sketching and part symbolizing.

The significant nature of the particular subject is not in question. The original method that the lecturer was describing formed the basis for the proof of the Poincaré conjecture that was confirmed by the offer of a Fields Medal to Gregori Perelman in 2006. It is

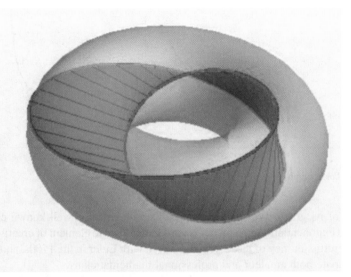

Fig. 3 A band in a torus

based on the concept of Ricci flow, concerning trajectories like the looping black curve in Fig. 1 but in a much more abstract space of tensors. As for the sketches, the reader can judge to what extent these are artistic, but there is no doubt that they reflect the creative nature of the geometric arguments involved and the extent to which contemporary mathematics naturally blurs the distinction between creativity and discovery.

The discussion, diagrammatic and otherwise, concerns the evolution of mathematical objects in time as described by differential equations, one of which is clearly visible in the top centre of Fig. 2. The first term in this equation is the so-called Ricci curvature tensor, R_{ij}, that features in Einstein's formulation of general relativity, but is here used to measure the distortion of a 3-dimensional object. At a more mundane level, it generalizes the description one can give of the convexity or concavity of a surface in space (like the ones in Figs. 1 and 3). The fourth temporal dimension is represented by the horizontal line.

Experts will recognize other main themes in Fig. 2, namely (bottom left) a simple Riemannian metric ds^2 in two isothermal variables x,y that represents the geometry of a surface, and the concept of a discrete quotient (*"this/G"*), which is used by mathematicians to describe an everyday object like a torus. A more concrete version of the latter is the opaque doughnut-shaped surface visibile in Fig. 3 housing a Möbius band whose red boundary is a knot on the torus surface.

2 Creativity Versus Discovery

As its name suggests, differential geometry combines the use of differential calculus with the description of geometrical objects. It grew out of more conventional analysis in which geometrical ideas abound, for example recognizing a function by

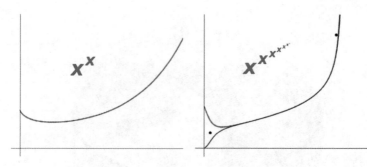

Fig. 4 Three graphs

means of its graph. We shall illustrate this idea with the well-known example of iterated exponentiation in which it is hard to deny some element of creativity amidst the scientific journey of discovery that began with Euler in the 1700s, and continues to fascinate both amateur and professional mathematicians.

We begin by plotting (in Fig. 4 left, though not quite to scale) what is visually one of the simplest looking functions, namely $y = x^x$, chosen for the very economy of notation with which one indicates the operation of exponentiation. The subsequent images on this and the next page refer to repeating this exponential process. By turning the handle just a few times, one obtains a sequence of graphs like the two in Fig. 4 right that combines the plots of eight x's (red) and nine x's (blue) arranged in so-called "power towers".

As we take more and more x's, one can experimentally verify a theorem of Euler, namely that if the tower is continued ad infinitum, it approaches a definite limit precisely when

$$e^{-e} \leq x \leq e^{1/e},$$

the extreme points of convergence left and right being indicated by black dots. A modern proof relies on the contraction mapping theorem and the continuity method. The limit function (that one can imagine extending between the black dots) is in fact the inverse function of $x^{1/x}$. (Graphically, the two functions are mirror images in the diagonal graph $y = x$, whilst experts will know that the inverse can be expressed in terms of the Lambert w function.) With the power of today's personal computers, it is easy to discover this phenomenon for oneself.

The appearance of a bifurcation around the left black dot in Fig. 4 right is inevitable since, as x tends to zero, the graph tends to 1 for an even number of x's, but to 0 for an odd number of x's. Euler tells us that the divergence occurs when x is about 0.03. After extending the graphs of Fig. 4 to the left of the vertical axis and into three dimensions with the aid of complex numbers, one soon discovers two trifurcation points, where the graph divides into three branches. The extended 3-dimensional graph is displayed in Fig. 5 (in which a brown line corresponds to the vertical axes of Fig. 4) whose horizontal scale has been exaggerated.

Fig. 5 Bifurcation leading to
two trifurcations

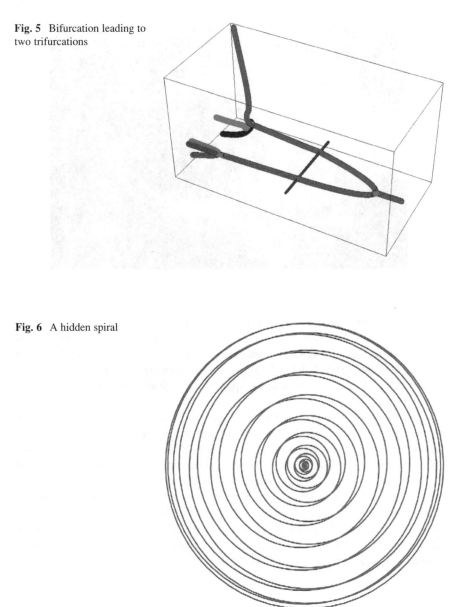

Fig. 6 A hidden spiral

In fact the two tripod legs on the left of Fig. 5 already house some surprises, like many slim helices that project to spirals of the type shown in Fig. 6. This one is thousands of times off the scale of Fig. 5, which illustrates period doubling and tripling familiar in other discrete dynamical processes.

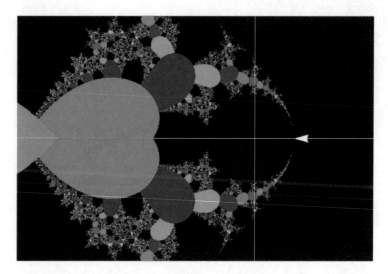

Fig. 7 Exponential convergence

Following the flow of the vessels beyond the left edge of the box in Fig. 5, we are soon be led to the onset of chaos. An incomplete fractal picture is illustrated as Fig. 7, in which the bifurcation point corresponds to the tip of the white arrow. It incorporates geometrical tell-tale signs of the convergence theory of iterated functions, such as the cardioid shape familiar from the analogous set of Mandelbrot.

One can think of Fig. 7 as the floor plan associated to a video game; at the bifurcation point one is abandoning Euler's interval of certain convergence for the duality inherent in the second stage that takes place in the first almost-circular chamber. The triple forking occurs at the next doorway at the centre of the overall black rectangle. Soon after that, quantum-like effects occur as gems that are to be plucked out of the air for extra points, though their presence goes unnoticed on the accompanying low-resolution images. Such gems include various spirals; the one in Fig. 6 could with hindsight be drawn with a simple mathematical formula using the techniques of [3], though it is actually the genuine object that was plotted by computer with exponential iteration.

3   Serret-Frenet Geometry

The previous section was designed to show how a simple mathematical idea can quickly lead to objects of aesthetic value. However, the main theme of the lecture concerned the construction of surfaces related to space curves and vector fields. We shall discuss the former in this section and the latter in the next. A space curve is formed by the trajectory of a particle (or a small flying insect) in space. As such it is described by a triple

$$(x(t), \, y(t), \, z(t))$$

of three functions of an independent variable t. We usually suppose that the curves are "smooth" (technically speaking, the three functions of t must be differentiable several times).

Whilst t can be thought of as time, it is better to make use of a different variable, denoted s, that represents arc-length measured along the curve. This is a natural parameter that is uniquely specified by the curve (independently of how it is traced) and a starting point (that we take to correspond to $s = 0$); if the curve were a piece of cotton, it would suffice to pull it tight and measure its length from the anchored starting point.

Mathematically it is known that any curve in space can be completely described by two functions of arc-length: the so-called *curvature* K (s) and *torsion* $\tau(s)$, traditionally indicated by Greek letters. Once these are assigned, they determine the curve up to a rigid motion: there is a unique one, once we assign an initial point and an initial direction for the curve to be tangent to. For example, the curve in Fig. 8 corresponds to

$$\kappa(s) = s \cos s, \qquad \tau(s) = \log s,$$

starting from $s = 1$ (the straighter part on the left). This figure illustrates the significance of curvature and torsion, which we explain next.

The curvature of a point in the plane measures the bending of the curve at that point. More precisely, it is equal to the rate of change of the angle between

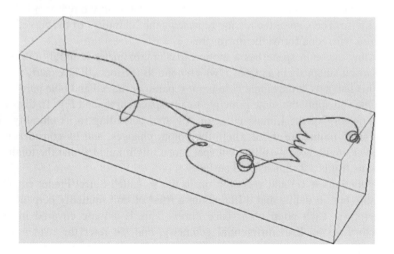

Fig. 8 A space curve with assigned curvature and torsion

the tangent line to the curve and a fixed direction. When the curve is a circle of radius r, the bending is the same at any point of the circle, and the curvature assumes the constant value $1/r$, In this sense, curvature equals the reciprocal of radius.

By taking plane sections of surfaces, one can discuss the curvature of a surface relative to any tangent direction and the maximum and minimum values of this function have special significance. Their product is the so-called Gaussian curvature that is generalized by the tensors of Ricci and Riemann, the basis for measuring the curvature of space-time in General Relativity. Positive Gaussian curvature at a point indicates that the surface is bending away entirely to one side of its tangent plane, and represents convexity at that point.

A similar definition applies to the curvature of a space curve. Once one has specified a point of the curve by assigning the value of s, then K (s) equals $1/r$, where r is the radius of the circle that best fits the curve at that point. A circle is a very special sort of space curve not just because its curvature is constant but because (as a consequence) it always lies in some plane. In general, the fit of a circle to a space curve at a given point can only be approximate, though one can prove that there is a best fit (the "osculating" or "kissing" circle). In Fig. 8, the curve has alternating straight and curly segments since $s \cos s$ is oscillatory and keeps returning to zero. But the amplitude of its peaks becomes greater and greater, so the curly interludes become more pronounced as one moves to the right.

The only setback for space curves is that there is no way of assigning a sign to the curvature function K which therefore (like the mass of an object) takes only non-negative values. By contrast, the torsion of a space curve has a definite sign that is best appreciated by considering a helix. As we look at such a curve, it is spiralling away from us in either a clockwise or an anticlockwise fashion. In Fig. 9, the red curve is spiralling away clockwise whether one looks from above or from below, but the blue curve (once when grasps its position relative to the red one) is fleeing anticlockwise. The sign of the torsion represents the "chirality" of the curve, which is reversed when one forms the mirror image.

Now any curve in space has a motion that approximates a spiral, unless it is moving instantaneously in a plane, at which point its torsion will be exactly zero. In Fig. 9, this happens at the single base point separating the red and blue parts of the curve. One can apply the same principle to any space curve, and Fig. 10 displays a single trajectory of the Lorenz attractor coloured according to its chirality. The "null-torsion" points are those where the colour changes, and by considering all trajectories simultaneously filling out space, one might imagine that the totality of such points froms a 2-dimensional surface.

The functions K (s) and $\tau(s)$ crop up in the so-called Serret-Frenet equations that enable one to define and differentiate a triad of unit mutually perpendicular unit vectors at each point of a space curve. This is a topic covered in every introductory textbook to differential geometry, and we refer the reader to, for example, [3–5].

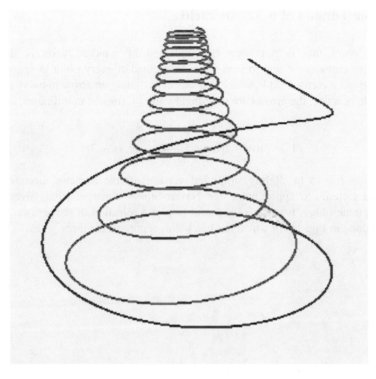

Fig. 9 The chirality of spirals

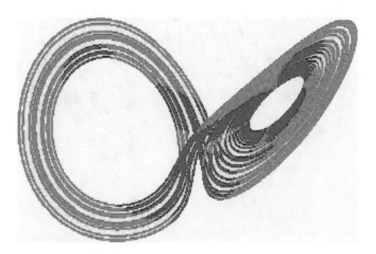

Fig. 10 Lorenz attractor coloured by torsion

4   The Torsion of a Vector Field

A key point that is not normally explained in standard texts is that the quantities curvature K and torsion τ can be defined at every point in space, once one is given a vector field V, which consists of assigning an arrow to every point in space. In practice, the arrows are specified by their Cartesian coordinates, as in the example

$$V = (10y - 10x, -xz + 28x - y, xy - 3z) \tag{1}$$

that gives rise to (a slightly simplified version of) the attractor discovered by Edward Lorenz, to approximate the Navier-Stokes equations that govern fluid flow in meterology. It is not very useful to plot a selection of the arrows, though this is done in Fig. 11, in which the black dots represent the three points

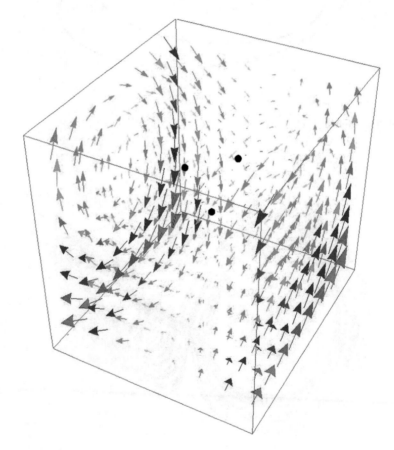

Fig. 11 A vector field of arrows

$$(0,0,0), \quad (9,9,27), \quad (-9,-9,27),$$

where the three components of V are simultaneously equal to zero.

Starting from any point (apart from a black dot), one can in theory follow the arrows and generate a unique trajectory. It is more interesting to visualize families of trajectories, like the one in Fig. 10. In this way, the vector field gives rise to a family of spaces curves that never touch one another. (The vector field is allowed to be zero at one or more points; at such points the arrow has zero length, no curve passes through the point, and strictly speaking neither K nor τ have a value there.) Whilst the typical trajectory is infinite in length, it is known that certain trajectories are actually closed curves or knots of varying complexity [1].

We can see from the definition of V on the previous page that a vector field is specified by giving three functions, each of three variables. It is an easy matter to compute the curvature and torsion by computer from an explicit knowledge of such functions. It requires computing two derivatives of the components of V, and then combining them in a highly non-linear fashion. Figure 12 is output from the program *Mathematica* expressing the torsion of the Lorenz vector field V. The details are unimportant, but it is worth noting that the equation itself is not at all elegant. Instead it is the definition or algorithm that leads to this output that is elegant.

If we set the quantity in Fig. 12 equal to zero, we obtain the "null-torsion" surface of Fig. 1 consisting of points where the torsion of V is zero. Actually, that is not quite true, there are other points with zero torsion not shown—namely those lying on the z-axis $x = 0 = y$, which is itself a trajectory of the Lorenz field. The surface is one component of the variety of degree 8 defined by the vanishing torsion. More vivid views of the null-torsion surface (with trajectories) can be seen in Figs. 13 and 14, and another in Fig. 15.

One can think of the Fig. 14 as representing a "stand" that could be placed on an office desk in order to properly display the more famous Lorenz curves. The second "upside-down" view betters displays the self-intersections of the surface, and a key feature of it, namely that it incorporates two "leaves" that are roughly planar (in

$$
\begin{aligned}
& x^6\, y^2 + x^6\, z^2 - 56\, x^6\, z + 784\, x^6 - 2\, x^5\, y z - 112\, x^5\, y + 20\, x^4\, y^2\, z - \\
& 558\, x^4\, y^2 + 20\, x^4\, z^3 - 1678\, x^4\, z^2 + 48944\, x^4\, z - 486864\, x^4 - 30\, x^3\, y^3\, z + \\
& 480\, x^3\, y^3 - 30\, x^3\, y z^3 + 2140\, x^3\, y z^2 - 53164\, x^3\, y z + 492912\, x^3\, y + \\
& 570\, x^2\, y^4 + 600\, x^2\, y^2\, z^2 - 29760\, x^2\, y^2\, z + 344817\, x^2\, y^2 - 30\, x^2\, z^4 + \\
& 2460\, x^2\, z^3 - 70623\, x^2\, z^2 + 666792\, x^2\, z - 300\, x\, y^5 - 300\, x\, y^3\, z^2 + \\
& 16870\, x\, y^3\, z - 244080\, x\, y^3 + 270\, x\, y z^3 - 13140\, x\, y z^2 + 214326\, x\, y z + \\
& 100\, y^4\, z + 13500\, y^4 - 300\, y^2\, z^3 + 15480\, y^2\, z^2 - 238140\, y^2\, z
\end{aligned}
$$

Fig. 12 A polynomial of degree eight

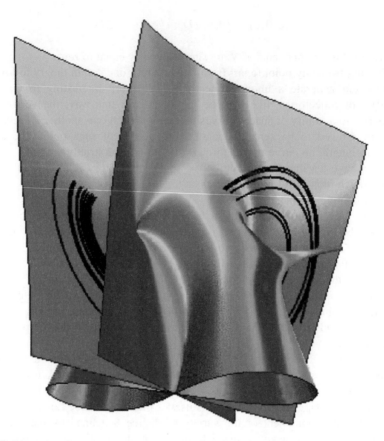

Fig. 13 Lorenz stand

theory these extend to infinity but of course we are only plotting part of the surface). One can understand this feature as follows. Roughly speaking, the butterfly attractor does itself fit into two planes that intersect at an acute angle (this is best seen in Fig. 10). If this were exactly true then all the points on the corresponding planar trajectories would have zero torsion (recall that τ measures the extent to which a curve does not lie in a plane).

The colours of Figs. 14 and 15 have a precise mathematical meaning. The spectrum (Red–Orange–Yellow–Green–Blue–Indigo–Violet) is used to indicate the angle with which the trajectories following the vector field emerge from the surface. Red indicates that they are tangent, so if the two leaves were exact planes they would be painted red all over. At the opposite extreme, violet represents points where the trajectories emerge at right angles (visible on subsequent images). Green would represent roughly a 45° angle of incidence between surface and curve.

Another aspect of the surface, more evident in the versions of Figs. 1 and 15, is the presence of cusps that form at two of the three points where the vector field

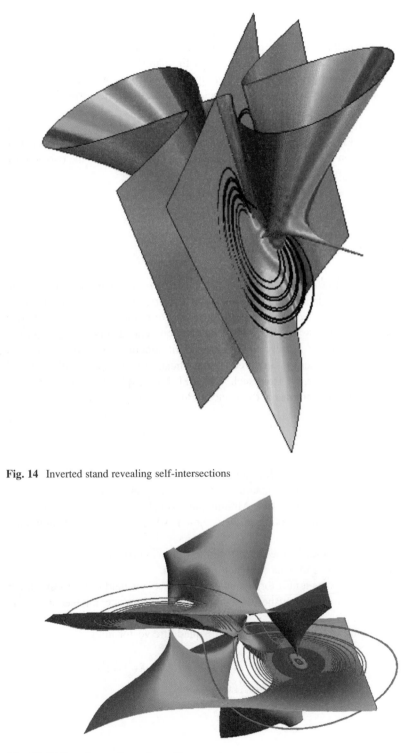

Fig. 14 Inverted stand revealing self-intersections

Fig. 15 Null-torsion surface coloured by curvature

is zero (physically, these can be thought of as equilibrium points of the dynamical system). Near these two points, the surface resembles a double cone (whose vertex is the respective point) separated by a plane. Actually, the surface is singular at all three such points, although the nature of the singularity at the origin is more complicated. What is perhaps more surprising is the presence of external spikes, though these point to the behaviour of other trajectories that are not shown.

Although the blue-green surface of Fig. 15 is also defined by setting the polynomial in Fig. 13 equal to zero, its colouring is related to the non-negative curvature function K. The latter becomes infinite as one approaches the two equilibrium points.

5   A Gallery of Null-Torsion Surfaces

Having mastered the manner in which the surfaces on the previous pages were constructed from the data defining the vector field V, it is an easy matter to apply the technique (and in practice, the computer program) to obtain other images. Above all, it is remarkable how much varied behaviour one encounters by restricting to quadratic vector fields, namely those component functions are polynomials of degree at most two. There is a degree of classification of the physically-relevant dynamical systems that arise from such fields in [2]. As explained there, one of the deceptively simplest examples is the vector field

$$W = (-y - z, x + \frac{1}{2}y, 2 + xz - 4z),$$

defined by the biochemist Otto Rössler (also a recent critic of the LHC). Notice that only the third component is quadratic—the first two are linear. Its associated trajectories form an analogue of a Möbius band (cf. Fig. 3) that is supported by its own version of the null-torsion surface, shown as Fig. 16 with a more autumnal colouring.

We shall briefly describe the images on the following page.

Figure 17 was chosen for its compact trajectories.

Figure 18 represents a null-torsion surface for a vector field close to the linear case, in which the torsion vanishes on a union of planes in space, intersecting at the origin.

The remaining two figures arise from the quadratic vector field

$$Q_a = (1 + yz + ax^2, 1 + zx + ay^2, 1 + xy + az^2)$$
$$= (1, 1, 1) + (yz, zx, xy) + a(x^2, y^2, z^2) \tag{2}$$

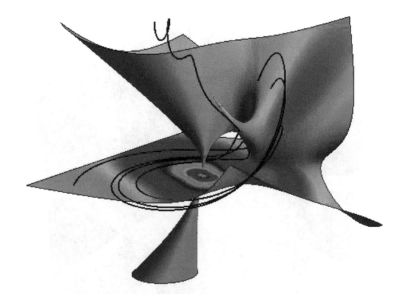

Fig. 16 Null torsion for the Rössler attractor

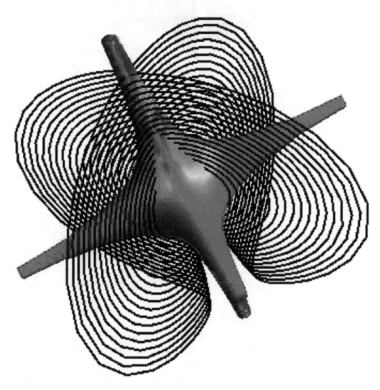

Fig. 17 Coiled trajectories

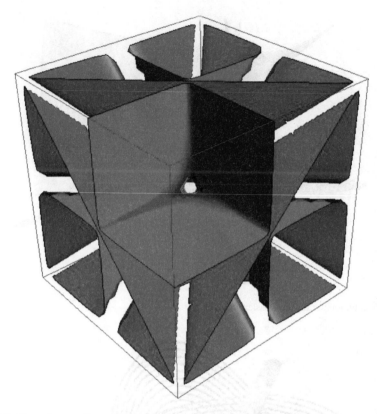

Fig. 18 Near-planar surface

for different values of the parameter a, Fig. 19 has $a = 1$ and Fig. 20 has $a = 1/4$. We explain below that the resulting vector fields have a threefold symmetry (this is more evident in Fig. 20).

On the next page, we display vanishing torsion arising from the vector field

$$(\sin y, \sin z, \sin x),$$

which is triply periodic since if we move a distance 2π in the x, y or z direction the value of the vector field remains the same. Figure 21 displays the associated null-torsion surface, coloured as in Figs. 13 and 14. Figure 22 is a view of one "cube" of the surface with some associated trajectories, which we see spiralling in the centre of the picture so as to arrive tangent to the red zone.

The vector fields giving rise to the previous images all possess a certain degree of symmetry, characterized by the action of some group of transformations that leave the equations invariant. For example, the Lorenz field V defined by (1) is unchanged when the signs of both x and y are reversed. This represents invariance by a 180° rotation around the z axis, which incidentally identifies two of the three

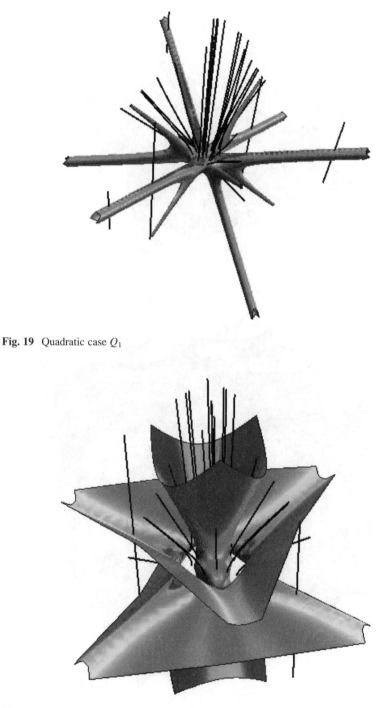

Fig. 19 Quadratic case Q_1

Fig. 20 Quadratic case $Q_{1/4}$

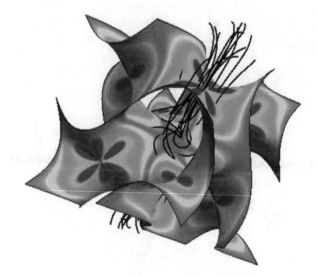

Fig. 21 A triply periodic surface

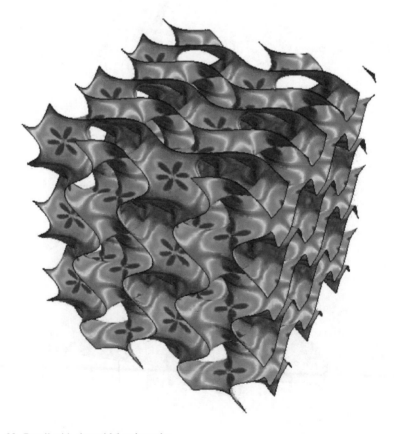

Fig. 22 Detail with sinusoidal trajectories

equilibrium points. For the sinusoidal figures, the group consists of translations defined by a lattice in space.

The symmetry realized by the vector field Q_a of (2) is threefold. To explain this, let ρ denote a rotation of 120° about the axis (1,1,1)—this is simply the linear transformation of $\mathbb{R}^3$ that cyclically permutes the coordinates:

$$\rho(x, y, z) = (y, z, x).$$

Then for each fixed a, the vector field Q_a has the property that (as a function $\mathbb{R}^3 \to \mathbb{R}^3$) it commutes with ρ:

$$Q_a(y, z, x) = \rho(Q_a(x, y, z)).$$

A consequence of this is that the "tri-rotation" ρ maps each trajectory of Q_a onto another trajectory.

Figures 23 and 24 arise from the vector field Q_a in (2) with $a = -1$. For this unique value of the constant a, the torsion is identically zero, and the

Fig. 23 Gas-ring necklace

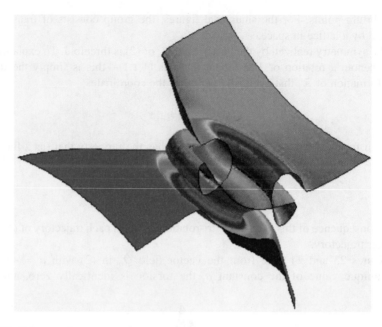

Fig. 24 Cut-away view

resulting vector field Q_{-1} has a continuous symmetry. Any one trajectory lies in a plane, but there is a rotational symmetry as this plane is allowed to rotate around the axis $(1,1,1)$. It turns out than one can factor out by the terms that makes the torsion zero, so as to plot the points where the torsion vanishes to second order. The result is the "gas-ring necklace" of Fig. 23. The smaller Fig. 24 is (despite a different colour scheme) a cut-away view to expose the neat self-intersection.

An obvious generalization of Q_{-1} is the vector field

$$(1 + yz - x^3, 1 + zx - y^3, 1 + xy - z^3). \tag{4}$$

This has cubic coefficients and the associated null-torsion surface is shown in Fig. 25 on the last page of this article.

Acknowledgments I first started using iterated exponentiation to illustrate bifurcation in a project for a Maple course I taught in Oxford in the 1990s. Remnants of this course are available on my homepage.

The study of space curves with assigned torsion and curvature is described in [3], and I used it as a student competition to find an interesting space curve by guessing simple candidate functions in a differential geometry course in Turin.

The vector field Q_a is the basis of the author's animation
Vector field Kaleidoscope on YouTube. This video contains frames of null-torsion surfaces for about 500 values of a and Figs. 19, 20 and 23 are instances of that.

Figure 25 arises from the vector field (4) with cubic components. As personal computing power increases, it will be easier to study more and more complicated vector fields and their associated surfaces.

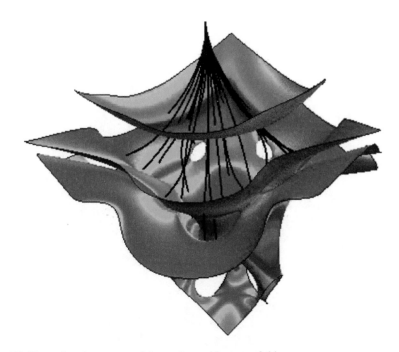

Fig. 25 Illustrating the geometry inherent in a cubic vector field

The ESMA conference taught me not just to better appreciate artistic aspects inherent in modern mathematics, and but also the importance of explaining to a wider public the underlying geometrical ideas I deal with on a daily basis. I found myself having to work on a number of deep mathematical problems to answer some of the questions that arose. In this respect, I wish to express gratitude to the participants, and in particular François Apery, Claude Bruter, Eugenia Emets, George Hart, and Jos Leys. Thanks are also due to Claude for his afternoon visits from Gometz to Bures encouraging me to finish the text.

References

1. Ghys, E.: Lorenz and modular flows, a visual introduction, www.ams.org/samplings/feature-column/fcarc-lorenz
2. Gilmore, C.R., Letellier, C.: The Symmetry of Chaos. Oxford University Press, London (2007)
3. Gray, A., Abbena, E., Salamon, S.: Modern Differential Geometry of Curves and Surfaces with Mathematica. CRC Press, FL (2006)
4. Pressley, A.: Elementary Differential Geometry, Springer Undergraduate Mathematics Series. Springer, Berlin (2001)
5. Struik, D.: Lectures on Classical Differential Geometry. Dover, NY (1988)

Pleasing Shapes for Topological Objects

John M. Sullivan

Abstract Topology is the study of deformable shapes; to draw a picture of a topological object one must choose a particular geometric shape. One strategy is to minimize a geometric energy, of the type that also arises in many physical situations. The energy minimizers or optimal shapes are also often aesthetically pleasing. This article first appeared in an Italian translation [Sullivan, Affascinanti forme per oggetti topologici, 145–156 (2011)].

1 Introduction

Topology, the branch of mathematics sometimes described as "rubber-sheet geometry", is the study of those properties of shapes that don't change under continuous deformations. As an example, the classification of surfaces in space says that each closed surface is topologically a sphere with a certain number of handles. A surface with one handle is called a torus, and might be an inner tube or a donut or a coffee cup (with a handle, of course): the indentation that actually holds the coffee doesn't matter topologically. Similarly a topological sphere might not be round: it could be a cube (or indeed any convex shape) or the surface of a cup with no handle.

Since there is so much freedom to deform a topological object, it is sometimes hard to know how to draw a picture of it. We might agree that the round sphere is the nicest example of a topological sphere—indeed it is the most symmetric. It is also the solution to many different geometric optimization problems. For instance, it can be characterized by its intrinsic geometry: it is the unique surface in space with constant (positive) Gauss curvature.

More physically, we can also consider the isoperimetric problem: among surfaces in space with a fixed surface area, which one encloses the most volume?

J.M. Sullivan
Technische Universität Berlin, Straße des 17. Juni 135, 10623 Berlin, Germany
e-mail: sullivan@math.tu-berlin.de

C. Bruter (ed.), *Mathematics and Modern Art*, Springer Proceedings in Mathematics 18, DOI 10.1007/978-3-642-24497-1_13, © Springer-Verlag Berlin Heidelberg 2012

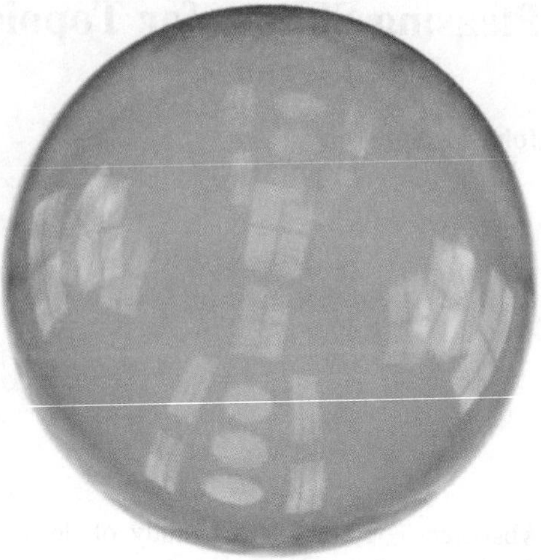

Fig. 1 A soap bubble
minimizes surface area for a
given enclosed volume, and
thus becomes a round sphere

Or equivalently: among surfaces enclosing a given volume, which uses the least surface area? Surface tension causes a soap bubble, as in Fig. 1, to almost instantly find the round sphere as the solution to this last problem. This answer was known to the ancient Greeks, but was first given a rigorous mathematical proof in the late 1800s.

2 Bubble Clusters and Foams

Clusters of two or more bubbles are not single smooth surfaces, but form examples of spaces that topologists call complexes: different sheets of surface joined together along curves. Again, the soap film seeks to minimize the area needed to enclose and separate the given volumes of the various bubbles. Certain basic results are known: each soap film in a cluster has constant mean curvature, and the films meet at constant angles along triple curves and at tetrahedral junctions. But bubble clusters and foams are still a source of many interesting open mathematical problems [15,19]. For instance, it was only around the turn of this century that mathematicians proved (see [12]) that the standard double bubble—made from spherical caps—beats all possible competitors as in Fig. 2.

For clusters of more than two bubbles, the area-minimizer is still unknown mathematically, but clusters of three or four bubbles can again be built out of spherical pieces as in Fig. 3, and these are conjectured minimizers.

Some bubble clusters have independent mathematical interest. For instance, the four-dimensional analog of the dodecahedron—known as the 120-cell or dodecaplex—can be radially projected to a three-sphere and then stereographically projected to ordinary space. The result, shown in Fig. 4, is a complicated and symmetric cluster of 119 bubbles [14].

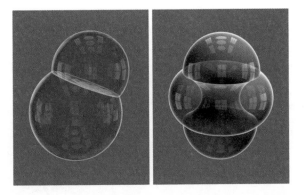

Fig. 2 A standard double bubble (*left*) consists of three spherical caps meeting at 120° dihedral angles along a circle. (The inner film curves slightly away from the smaller, higher pressure bubble.) It was relatively easy for mathematicians to show that the minimizing double bubble must have rotational symmetry, but then it was hard to rule out strange configurations (*right*) where one bubble forms a belt around the other, or even cases where the bubbles have several disjoint components

Fig. 3 The standard triple bubble is also built from pieces of round spheres meeting along circular arcs. Using Möbius transformations, we can make a version of this cluster with any desired triple of volumes. These are conjectured to be the optimal triple bubbles, but this has not yet been proven (except in 2D)

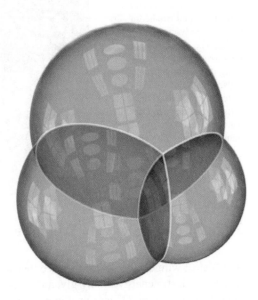

A foam—as found say in shaving cream or in the kitchen sink when doing the dishes—is like a cluster of many bubbles. Mathematically it is easiest to consider infinite, triply-periodic foams that fill all of space. Lord Kelvin [21] asked for the least-area foam whose cells all have equal volume. His conjectured symmetric solution remained unbeaten until more than a century after it was proposed (Fig. 5): the Weaire-Phelan foam [23] mixes two shapes of cells and ends up with less average surface area per bubble ([11] and reprinted in [22]).

Fig. 4 This cluster of 119
bubbles is the stereographic
projection of the dodecaplex
or 120-cell, a regular polytope
in four dimensions. One of its
120 dodecahedral cells
projects to the infinite outside
region. The others are
arranged in seven symmetric
layers around a tiny central
bubble

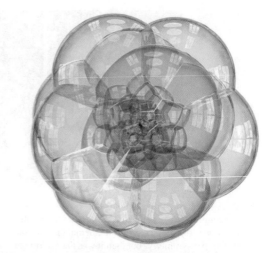

Fig. 5 The Kelvin foam (*left*)
fills space with congruent
cells, truncated octahedra
tiled in a body-centered cubic
lattice. The Weaire-Phelan
foam (*right*) saves area by
using two different shapes of
cells, with equal volumes but
different pressures. These
cells fit together in a crystal
pattern called A15, found in
transition-metal alloys

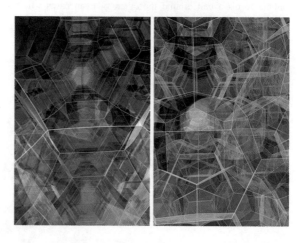

3   Sphere Eversions via Willmore Energy

A stiff steel wire can be bent into space, but will spring back to its original straight
shape. Its elastic energy in any given configuration is proportional to the integral
of curvature squared. (This is analogous to Hooke's law that the energy of a spring
is proportional to displacement squared.) The corresponding elastic bending energy
for surfaces has several forms which are equivalent by the Gauss–Bonnet theorem:
most common is the Willmore energy (see [13, 24]), the integral W of mean
curvature squared. Physically, many bilayer surfaces, such as biological cell
membranes, seem to minimize this energy. For instance, minimizing W while fixing
area and enclosed volume can lead to shapes like that of a red blood cell (Fig. 6).

Mathematically, it is also interesting to consider W for immersed surfaces, that
is, for surfaces which are allowed to self-intersect, but which have to stay smooth,

Fig. 6 This shape was obtained by minimizing the Willmore energy for fixed values of surface area and of enclosed volume. Cell membranes probably minimize this same energy, and indeed this picture is reminiscent of a red blood cell

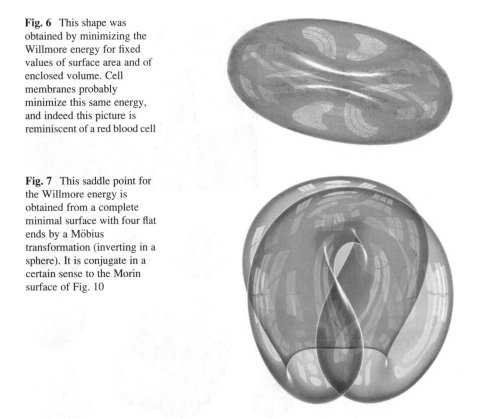

Fig. 7 This saddle point for the Willmore energy is obtained from a complete minimal surface with four flat ends by a Möbius transformation (inverting in a sphere). It is conjugate in a certain sense to the Morin surface of Fig. 10

with no creases, corners or rips. The sphere in Fig. 7 is immersed in a complicated way—making it hard to recognize as a sphere—but this shape is a stationary point (a saddle point) for the Willmore energy.

Immersed surfaces are the key to defining a sphere eversion (see [16]). To turn a sphere inside out physically, we have to cut a hole in the sphere, pull the rest of the surface through the hole (as when turning a sock inside out) and then finally patch the hole. Mathematically, the interesting problem is to do this without a hole: the surface must be an entire smooth sphere at all times, but it is allowed to be immersed with self-intersections. One sheet of surface can pass through another (like a ghost through a wall) without affecting the integrity of the sphere. But again, no pinching, creasing or ripping is allowed. After Smale proved abstractly that a sphere eversion was possible, other mathematicians took years to find the first explicit eversions.

One strategy is to use a projective plane, immersed as Boy's surface, at the halfway stage. That is, in the middle of the eversion, we immerse the sphere in such a way that each pair of antipodal points maps to a single point in space: two sheets of surface always lie exactly on top of each other. To perform the eversion, we pull

Fig. 8 Tom Willmore at the
mathematical research
institute in Oberwolfach in
the year 1998, standing next
to a metal sculpture of a
Boy's surface that minimizes
Willmore energy, made out of
ribbons of flat steel plate that
have been riveted together

Fig. 9 The threefold
minimax eversion starts with
the red sphere at the upper left
and moves clockwise. The
large center image and the
two images below it are near
the halfway stage, where we
have a double cover of the
Willmore Boy's surface

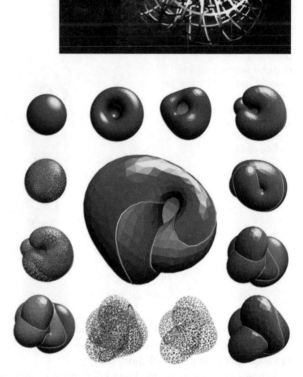

the sheets apart and simplify the result until it is a round sphere. The idea of the *minimax eversions* [7] is to do the simplification automatically by minimizing the bending energy W. This results in shapes which are more pleasantly rounded than in the topologically equivalent eversions designed earlier by hand. The halfway stage in a minimax eversion is a saddle point for W, for instance the Willmore-minimizing Boy's surface, which has also been depicted in a large metal sculpture (Fig. 8) at the mathematical research institute in Oberwolfach, Germany [10].

We simulated the minimax eversions numerically, and the resulting animations were featured in our video "The Optiverse" [20], premiered at the 1998 International Congress of Mathematicians in Berlin. The threefold eversion, passing through the Willmore Boy's surface, is shown in Fig. 9.

The simpler twofold minimax eversion uses a Willmore-minimizing Morin surface (Fig. 10) at the halfway stage. This immersed sphere has four lobes, and a 90° rotation interchanges the inside and outside. The surface is named after the

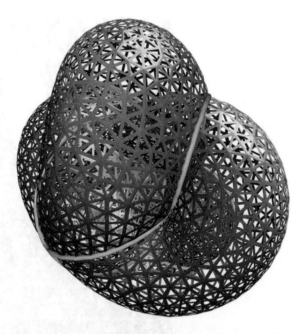

Fig. 10 A Morin surface is an immersed sphere with four lobes interchanged by rotational symmetry, two showing the inside surface (*blue*) and two showing the outside (*red*). It was described by Morin as the halfway stage of the simplest possible sphere eversion. This picture shows a Willmore-minimizing Morin surface; the white tubes highlight the self-intersection curves

Fig. 11 Bernard Morin, at a conference on Art and Mathematics in Maubeuge, France in the year 2000, exploring the geometry of a Willmore-minimizing Morin surface, built on a stereolithographic printer directly from the computer data

Fig. 12 We simulated the
minimax eversions on the
computer using polyhedral
surfaces with thousands of
triangular faces. To visualize
the intermediate stages of an
eversion, we can remove the
middle of each triangle to
look through the surface, and
emphasize the double curves
of self-intersection with white
tubes

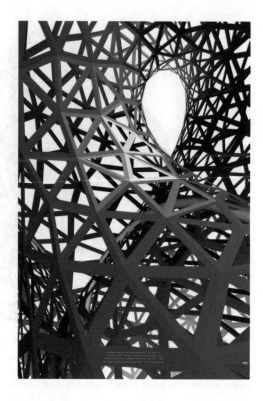

blind French mathematician Bernard Morin, who described it topologically for one
of the first explicit sphere eversions [6]. To show him what our Willmore-
minimizing version of the Morin surface looks like, we presented him with a
model created on a 3D-printer directly from the computer data, and he could
explore it with his fingers as in Fig. 11.

The twofold minimax eversion agrees topologically with Morin's eversion, and
because of its relative simplicity, we can understand it well in "The Optiverse" by
flying through the intermediate stages, with views like Fig. 12.

With a team led by Stan Wagon, we also carved a large Morin surface (though
not the Willmore version) out of snow (Fig. 13) at the 2004 International Snow
Sculpture Championships. The following year, we competed again, sculpting a
mathematical knot.

4   Tight Knots

Knot theory is the branch of mathematics concerned with giving a topological
classification of knots—simple closed space curves—by considering which curves
can be deformed into each other without crossings. Geometric knot theory looks at
connections between the topological complexity of a knot and the geometric

Fig. 13 This Morin surface (*left*), carved out of snow at 4 m scale, won an honorable mention for "Most Ambitious Piece" at the International Snow Sculpture Championships in Boulder, Colorado in January 2004. One year later, our team turned to knots as another source of interesting mathematical sculpture, carving a twofold cabling of a trefoil knot (*right*)

Fig. 14 The tight configuration of the Borromean rings has pyritohedral symmetry, with each component being a piecewise smooth planar curve described in part by elliptic integrals. These renderings in different styles are from the video "The Borromean Rings" [8, 9], premiered at the 2006 International Congress of Mathematicians in Madrid

complexity of any space curve realizing that knot type. One idea, with possible relevance for physical knots tied in rope, is to consider tight knots and links, made out of rope of fixed circular cross-section, pulled tight to minimize the length of rope needed. Although some basic theory for this problem is known [3], only a few tight links—where each component is a planar curve as in the Borromean rings (Fig. 14)—have been described explicitly [4].

This tight configuration of the Borromean rings, viewed along the axis of threefold symmetry, has been selected as the logo (Fig. 15) of the International Mathematical Union (IMU).

Some knots break their symmetry when pulled tight, or when minimizing other geometric knot energies. For instance, the (3,4)-torus knot has configurations with perfect three or fourfold symmetry. But if we minimize a certain repulsive-charge knot energy related to the Coulomb potential, this symmetry gets broken as in Fig. 16. The knot then reveals itself as an interweaving of two trefoil knots.

Fig. 15 The new logo of the
IMU (*left*), chosen in an
international competition, is a
threefold symmetric view of
the tight Borromean rings

Fig. 16 The (3,4)-torus knot
in an energy-minimizing
configuration (*right*) breaks
its symmetry and exhibits
itself as two interlocked
overhand knots: this has been
called the "true lover's knot"

Fig. 17 A stereo view of the (presumably) tight configuration of the Turk's head knot 8_{18}, rendered by Charles Gunn. The *middle* image is for the right eye: to see the stereo effect, view the left pair wall-eyed or the right pair cross-eyed

While no tight knot of a single component is known explicitly, numerical simulations show some cases (like the Turk's head knot in Fig. 17) keeping their symmetry, with pieces geometrically similar to those found in the Borromean rings.

5 Mathematical Visualization and Art

We have looked at a number of geometric optimization problems, and have seen how their solutions often exhibit graceful shapes. But if we want to depict them— either as three-dimensional sculptures or as two-dimensional images—there are challenges because of their complexity. Mathematically interesting curves and surfaces often have lots of hidden interior structure: the middle stages of a sphere eversion have complicated self-intersections, foams fill space with touching bubbles, and the various strands in a tight knot or link push up against each other.

Many of the figures in this article depict transparent surfaces. These are computer graphics renderings made with Pixar's RenderMan software, using the custom-made shader for soap films ([1] and Color Plate C, reprinted in [5]) that I programmed using the Fresnel laws of thin-film optics. Notably, the transparency of a soap film is much lower when viewed obliquely; this feature is missing from most simple computer graphics algorithms for transparency. Sometimes we have artificially darkened the transparent surfaces, in order to better show which sheets of the surface are in front of the others.

Often, the geometry depicted comes from numerical simulations using Brakke's Surface Evolver [2]. Although this program was originally designed to minimize area, as in bubble cluster problems, the Evolver has been extended to minimize many other geometric energies, including the elastic bending energies and knot energies we have discussed. For most of the situations described here, the mathematical theory lags behind the numerical simulations, and there are interesting open problems related to proving that these pictures are accurate. The interplay between numerical simulations (with visualizations) on the one hand and rigorous proofs on the other is what allows progress on both fronts.

Optimization principles can also be used to design mathematical artwork that goes beyond pictures created primarily for visualization purposes. My mathematical sculptures *Minimal Flower 3* and *Minimal Flower 4*, shown in Fig. 18, combine the ideas of minimal surfaces and knots, and were inspired by work of Brent Collins. They were first exhibited together at the Institut Henri Poincaré in Paris in 2010.

The first step in creating them [17] is to design a knotted boundary curve with the desired three or fourfold symmetry and span it with a crude surface having the

Fig. 18 The sculptures *Minimal Flower 3* and *4* depict minimal surfaces spanning knotted boundary curves. This picture was taken at the first meeting of the European Society for Mathematics and Art in Paris in July 2010

correct topology. Then the Evolver can be used to minimize the area of this spanning surface until it has the geometry of a soap film. Here we have to be careful to maintain the symmetry, as the desired surfaces are unstable soap films, not least-area surfaces. The aesthetic effect is improved if we actually work in the conformal ball model of hyperbolic space, accentuating the U-shaped cross-section of the outer loops. Finally, the mathematical surface has to be given a tapered thickness to create the physical sculpture.

References

1. Almgren, F.J. Jr., Sullivan, M.: Visualization of soap bubble geometries. Leonardo **24**(3/4), 267–271 (1992)
2. Brakke, K.A.: The surface evolver. Exp. Math. **1**(2), 141–165 (1992)
3. Cantarella, J., Kusner, R.B., Sullivan, J.M.: On the minimum ropelength of knots and links. Inventiones Math. **150**(2), 257–286 (2002), arXiv:math.GT/0103224
4. Cantarella, J., Fu, J., Kusner, R., Sullivan, J.M., Wrinkle, N.: Criticality for the Gehring link problem. Geom. Topology **10**, 2055–2115 (2006), arXiv.org/math.DG/0402212
5. Emmer, M. (ed.): The Visual Mind: Art and Mathematics. MIT, Cambridge (1993)
6. Francis, G., Morin, B.: Arnold Shapiro's eversion of the sphere. Math. Intell. **2**, 200–203 (1979)
7. Francis, G., Sullivan, J.M., Kusner, R.B., Brakke, K.A., Hartman, C., Chappell, G.: The Minimax Sphere Eversion. In: Hege, H.-C., Polthier, K.(eds.) Visualization and Mathematics, pp. 3–20. Springer, Heidelberg (1997)
8. Gunn, C., Sullivan, J.M.: The Borromean rings: a new logo for the IMU. In: Polthier, K., Aigner, M., Apostol, T.M., Emmer, M., Hege, C.-H., Weinberg, U. (eds.) MathFilm Festival. Springer, Berlin (2008); 5-minute video
9. Gunn, C., Sullivan, J.M.: The Borromean rings: a video about the new IMU logo. Bridges Proceedings (Leeuwarden), pp. 63–70 (2008)
10. Karcher, H., Pinkall, U.: Die Boysche Fläche in Oberwolfach. Mitteilungen der DMV **97**(1), 45–47 (1997)
11. Kusner, R., Sullivan, J.M.: Comparing the Weaire-Phelan equal-volume foam to Kelvin's foam. Forma **11**(3), 233–242 (1996)
12. Morgan, F.: Proof of the double bubble conjecture. Am. Math. Monthly **108**(3), 193–205 (2001)
13. Pinkall, U., Sterling, I.: Willmore surfaces. Math. Intell. **9**(2), 38–43 (1987)
14. Sullivan, J.M.: Generating and rendering four-dimensional polytopes. Math. J. **1**(3), 76–85 (1991)
15. Sullivan, J.M.: The geometry of bubbles and foams. In: Rivier, N., Sadoc, J.-F. (eds.) Foams and Emulsions. NATO Advanced Science Institute Series E: Applied Sciences, vol. 354, pp. 379–402. Kluwer, Dordrecht (1998)
16. Sullivan, J.M.: The Optiverse and other sphere eversions. Bridges Proceedings (Winfield), pp. 265–274 (1999), arXiv:math.GT/9905020
17. Sullivan, J.M.: Minimal flowers. Bridges Proceedings (Pécs), pp. 395–398 (2010)
18. Sullivan, J.M.: Affascinanti forme per oggetti topologici. In: Emmer, M. (ed.) Matematica e cultura 2011, pp. 145–156. Springer, Italia (2011)
19. Sullivan, J.M., Morgan, F. (eds.): Open problems in soap bubble geometry. Int. J. Math. **7**(6), 833–842 (1996)

20. Sullivan, J.M., Francis, G., Levy, S.: The Optiverse. In: Hege, H.-C., Polthier, K. (eds.) VideoMath Festival at ICM'98, p. 16. Springer, Berlin (1998); plus 7-minute video, torus. math.uiuc.edu/optiverse/
21. Thompson, W. (Lord Kelvin), On the division of space with minimum partitional area. Philos. Mag. **24**, 503–514 (1887), also published in Acta Math. **11**, 121–134
22. Weaire, D. (ed.): The Kelvin Problem. Taylor & Francis, London (1997)
23. Weaire, D., Phelan, R.: A counter-example to Kelvin's conjecture on minimal surfaces. Phil. Mag. Lett. **69**(2), 107–110 (1994)
24. Willmore, T.J.: A survey on Willmore immersions. In: Geometry and Topology of Submanifolds, IV (Leuven, 1991), pp. 11–16. World Scientific, Singapore (1992)

Rhombopolyclonic Polygonal Rosettes Theory

François Tard

Abstract The division of a regular polygon with an even numbers of vertices into a whole number of "isoperimetric" rhombuses (equal sides and different angles) is possible. Elementary reasoning leads to a general theory, which offers many possibilities of plastic applications.

1 Introduction

The following natural idea occurred to me. I started by drawing a fractal figure, dividing up a regular dodecagon into triangles (Fig. 1).

Colouring this drawing, I obtained a picture that is reminiscent of the diffraction of light (Fig. 2).

However, I found this picture on the cover of a publication dedicated to the golden ration. Therefore, I took no further interest in this particular example. I asked myself what would happen if rather than divide the inside of a polygon into triangles I instead divided it into rhombuses.

Let p denote the length of the of an edge of the polygon and c denote the length of the edge of a rhombus.

I first tried to divide the polygon into successive rings of isoperimetric rhombuses starting from the periphery and extending towards the centre (Figs. 3 and 4).

In either case, it is not clear what is going to happen when we approach the centre of the polygon.

The idea occurred to me to invert the process, starting from the centre of the polygon.

F. Tard (✉)
1, rue Pierre Mille, 75015 Paris, France
e-mail: tard.francois@wanado.fr

C. Bruter (ed.), *Mathematics and Modern Art*, Springer Proceedings in Mathematics 18,
DOI 10.1007/978-3-642-24497-1_14, © Springer-Verlag Berlin Heidelberg 2012

Fig. 1 Fractal octodecagon

Fig. 2 "Diffraction of light"

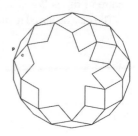

Fig. 3 Starting from the periphery with c = p

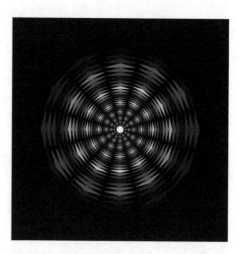

Fig. 4 Starting from the
periphery with c = p/2

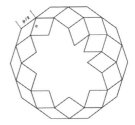

2   Construction of Rhomboid Polygons from a Regular "Polyclon"

Given a rhombus whose vertices are (A,B,C,D), whose edges are (AB), (BC), (CD), (DA), its diagonals are (A,C) and (B,D). The angle between the edges (AB) and (AD) will be denoted by α, the angle between the edges (BA) and (BC) by β ($\beta = \pi - \alpha$)·

We denote by R(c, α) a rhombus whose edge length is c, and which has an interior angle of α·

n being an integer, we suppose $n \geq 3$, $n \neq 4$.

Definitions:
1. Two rhombuses R(c, α) and R(c′, α′) are *isoperimetric* if c = c′.
2. Given a regular convex n-polygon (n is its number of edges), a *n-ring* of isoperimetric rhombuses over the given n-polygon consists in the data of n isoperimetric rhombuses whose one diagonal of each one is an edge of the polygon.
3. If the centre O of the polygon is a common vertex to all the rhombuses, the n-ring will be said to be of *rank* 1, the polygon will be called a *polyclon*, or a *basic polygon*, or a *rank 1 polygon*.

Here is an example of a rank 1 9-ring (Fig. 5).
Starting from a rank 1 n-polygon, it is easy to construct a series of new polygons.

Example 1: Let us start with the rank 1 9-ring of Fig. 5. We construct a second rank 9-ring in the following way (see Fig. 6): the n = 9 vertices of this ring which are opposite to the centre O of the basic polygon define a new regular n-polygon, which is a rank 2 9-polygon over the basic polygon. The n = 9 vertices of the rank 2 9-polygon will be named V_{2i} (I = 1, 2, ..., n) and will called the vertices of second rank. The n = 9 edges of the second rank regular n-polygon are the diagonals of n = 9 new rhombuses which are isoperimetric with the previous ones. They form a rank 2 n = 9-ring of rhombuses.

The n = 9 vertices V_{1i} (I = 1, 2, ..., n) of the rhombuses of the first ring which are not opposite to O, will be called *first rank vertices*.

If α_{1n} is the angle of a rhombus in O, the angle α_{2n} in V_{1i} of a rhombus from the second ring is given by $\alpha_{2n} = 2\alpha_{1n}$ (Fig. 7).

Fig. 5 Enneaclon (a rank
1 9-ring)

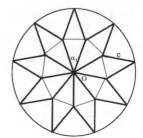

Fig. 6 Beginning of the
construction of a rank 2 9-ring
of rhombuses

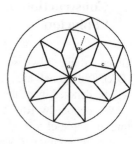

Fig. 7 Angles on two
successive rhombi

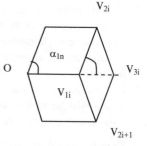

Note that $\alpha_{1n} = 2\,\pi/n$, but since the previous relation does not depend on n, we shall drop the index n in the next relations between angles.

Constructing by the same process a third rank n = 9-ring of rhombuses, we observe that we get a convex regular $2 \times n = 2 \times 9 = 18$-polygon (Fig. 8).

Let α_3 be the angle in V_{2i} of a rhombus from the third ring. From the relations:

$$2(\pi - \alpha_2) + \alpha_1 + \alpha_3 = 2\pi$$

$$\alpha_2 = 2\alpha_1$$

we get:

$$\alpha_3 = 3\alpha_1$$

Fig. 8 Octodecagon
obtained from an enneaclon

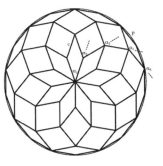

Suppose that now we add a fourth ring of rhombuses according to the previous process: the n vertices V_{3i} on the rhombuses which are opposite to the vertices V_{1i} define a rank 3 regular n-polygon from which a rank 3 of n rhombuses is constructed.

From similar relations to the previous ones:

$$2(\pi - \alpha_3) + \alpha_2 + \alpha_4 = 2\pi$$

$$\alpha_3 = 3\alpha_1$$

$$\alpha_2 = 2\alpha_1$$

one gets:

$$\alpha_4 = 4\alpha_1$$

More generally, by trivial induction, since $\alpha_1 = 2\pi/n$,

Lemma 1:
$$\alpha_k = k\alpha_1 = 2k\pi/n.$$

In the example, $\alpha_1 = 2\pi/9$: when $k = 4$, $\alpha_4 = 8\ \pi/9$, while $\alpha_5 = 10\ \pi/9$. Since any interior angle of a rhombus is less than π, one cannot add in the present case a new ring of rhombuses, so that the process has to stop with the construction of the fourth ring of rhombuses.

Definitions: The adding of a ring of rhombuses to a basic polygon will be called a *partial expansion* of the polygon with rhombuses. When the process of creating such new rings has to stop, we shall say that the obtained polygon is the *complete expansion* of the basic polygon. It will be called a *rhomboid polygon*.

Note that any expansion generates a polygon with the same number 2n of edges.

Fig. 9 Decagon arising from
a decagon

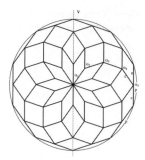

Corollary 1: The complete expansion of a basic regular n-polygon with rhombuses is a regular 2n-polygon consisting of r = [n/2] − 1 rings of rhombuses.

Proof. Since $\alpha_k = k\alpha_1 = 2k\pi/n < \pi$, then k < n/2, so that its maximum value is [n/2] − 1, where [n/2] is the integer part of n/2.

Corollary 2: The former complete expansion is a convex polygon.

Proof. The last two series of vertices of higher rank are the V_{ri} and the $V_{r+1,i}$ vertices. To add a new rhombus, we have to catch one V_{ri} vertex and two adjacent $V_{r+1,i}$ vertices. If the polygon was not convex in V_{ri}, then it would be possible to add the new rhombus. But since that is not possible, then the polygon has to be convex in V_{ri}. vex by construction at any $V_{r+1,i}$. Thus the complete expansion is convex at any vertex of its boundary.

Let us now look at the special case *when n = 2ρ:* In that case, on one hand r = ρ − 1. On the other hand, let us consider any V_{ri} vertex, and compute the sum of the angles of the three rhombuses having this vertex in common; setting $\alpha_1 = \alpha$, this sum is:
$$2(\pi - r\alpha) + (r - 1)\alpha = 2\pi - (r + 1)\alpha = \pi$$

since

$$\alpha = 2\pi/n = \pi/(r + 1)$$

Then:

Proposition: Whenever n is even, the boundary of the complete expansion is a regular n-polygon whose edge has length 2c (Fig. 9).

When besides *n = 6m*, then for k = 3m/2, the k-ring is made of squares since $k.\alpha = 3m/2(2\pi/6m) = \pi/2$

Such a situation also occurs for the m-ring when *n = 4m*.

3 Dividing a Regular n-Polygon, n Being an Even Integer

The following statement is a direct consequence of the results of the previous paragraph:

Theorem 1. Given any regular polygon with an even number N of vertices, there exists an infinite number of ways to divide it into isoperimetric rhombuses. Any given decomposition gives rise to an infinite associated series parametrized by the integers s^2 where $s \in \mathbf{N}$.

Proof. Let $N = 2n$ be the number of vertices of the polygon, p the length of an edge. The angle $\alpha = 2\pi/n = \pi/\rho$ and the value of p allow to define the main process of decomposition. There are two cases according to the parity of ρ.

When ρ is odd, we can construct $r = [\rho]\, n - 1$ rings of $R(c, \alpha)$ rhombuses with $c = p$, by starting from the centre.

When ρ is even, we can construct $r = \rho - 1$ rings of $R(c, \alpha)$ rhombuses with $c = p/2$ by starting from the centre.

Let $K = r.n$ the numbers of rhombuses constructed for each case. It is now possible to divide each rhombus into s^2 isoperimetric rhombuses whose edge has length c/s where s is any integer.

4 Picking Up a Rhombo-Polygonal Rosette on the Sphere

We shall only quote the main result:

Theorem 2: Let R be a rhombo-polygonal rosette generated by a rank 1 n-ring where n is even. The successive rings of rhombuses on the plane are inscribed into circles which are the projections of parallels on the 2-sphere with the same angular distance (Fig. 10).

Here is an image made for me by Los Leys: Fig. 11

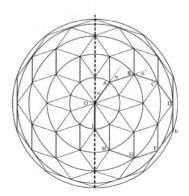

Fig. 10 A rhombo-
10-polygonal rosette

Fig. 11 Picking up on the sphere of a rank 1 18-ring

Fig. 12 Durer's rosette

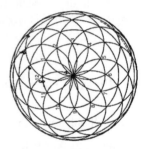

5   Historical Remarks and New Works

Though the past literature does not seem to refer to the facts shown in the previous paragraphs, it is interesting to recall preliminary works by Dürer and Képler respectively on rosettes and on rhombuses.

When the number of vertices increases, any rhombopolyclonic rosette looks more and more like some Durer's drawing (Fig. 12) [1].

In that figure, we find arcs of circles instead of straight lines segments, and vertices of rhombus take place at the intersections of circles. Compare with the following figure: Fig. 13.

Képler, studying the "thombic figure in the alveoles of bees" [2], got interested in constructing rhomboedric figures on the plane or in the 3-space. Here is a "pentarhomboclonic decagonal rosette" he has drawn (Fig. 14).

We shall conclude by showing two recent images: Figs. 15 and 16.

Fig. 13 Big blue rosette

Fig. 14 Képler's drawing

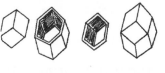

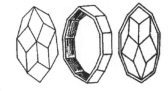

figure 8.

Construction   des deux Rhomboèdres.

Fig. 15 Octorhomboclonic
octagonal rosette

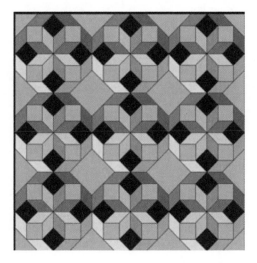

Fig. 16 Cyclic time snake

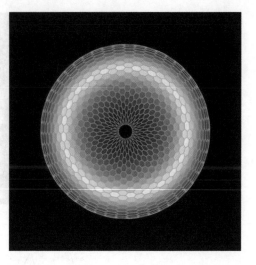

Acknowledgment I would like to thank Claude Bruter for the rewriting and the formatting of my text, Mike Field as well for his linguistic improvement.

References

1. Dürer, A.: Géométrie (présentation et traduction J. Peiffer). Seuil, Paris (1995)
2. Kepler, J.: L'Etrenne ou la neige sexangulaire (traduction et critique R. Halleux). CNRS-Vrin, Paris (1975)

Index

A
Attachment, 32, 44
Attractor, Lorenz, 131, 138, 139

B
Blowing-up, 40
Boy surface, 96, 97
   3-symmetry, 96
   5-symmetry, 96, 97
   wire-model, 18, 22

C
Coil, turn of, 111
Cordovan
   polygon, 117, 118, 120, 127, 128
   proportion, 117–129
Curvature, 93, 97–101, 112, 133, 137, 138,
      140, 141, 143, 144, 150, 153,
      154, 156
   of a curve, 137, 138
   Gauss, 105, 112
   Geodesic, 25
   mean, 154, 156
Cutting, 43, 44
Cyclids, 7–8, 20, 21, 23, 25
   Dupin, 14, 20, 21
   one sided, 20, 25
   ring parabolic, 23

D
Dandelin model, 24
Dini helicoid, 98
Dodecahedron, Poinsot great, 20–22

E
Etruscean Venus, 25, 26
Euler's formula, 106–107
Eversion, sphere, 156–160, 163

F
Folding, 37, 38, 43, 44
Four-dim polytope, 154, 156
Fractal, 6–8, 131, 136
Functions
   bi-periodic, 85, 101–103
   Jacobi, 101–102
   Weierstraß, 88, 89102–103

G
Genus, 106

H
Hyperbolic, 5
   geometry, 69–77
   plane, 69, 70
   surface, 2–8, 86–101, 103

I
Identification, 44
Illusion
   Aitken wheel, 60
   Ebbinghaus, 60–61
   Hering, 53, 54, 60
   Hermann grid, 60
   Wundt, 53, 60

Inflation, 35–37, 39–43
    regular, 39, 42–43
    singular, 35–37, 39–42
Isoperimetric problem, 153
Isoperimetric rhombus, 167, 169,
        172, 173

J
Jacobi functions, 101

K
Klein Bottle
    double, 95
    triple, 96
Knot, 105–115, 160–163
    Borromean link, 161, 162
    cyclic, 110–115
    trefoil, 108, 109, 161
    true-lover, 162
    Turk's head, 110, 162
Kuen surface, 3–5, 98–100

M
Minimal surface, 85–94
    Bonnet, 93
    Catalan, 89–90, 93
    Enneper, 88, 91
    Henneberg, 93–94
    Jeener, 90–91
Môbius band, 44, 112, 113, 133, 144
Monge formula, 89
Morin surface, 18, 98, 157–161
    wire model, 18, 22

N
Nodus, 111–113, 115

P
Petal loop, 110, 111
Pinching, 37–40, 44

Poincaré
    disk model, 71–74, 76
    half plane model, 69–76
Pseudo-sphere, 5, 98, 99

R
Rhomboic polygon, 167–176
Rosette, 167–176

S
Sierpinski carpet, 54, 55
Sievert surface, 25, 26, 100, 101
Singularity, 31, 35, 38, 41
    antibubbling, 33, 37, 38, 41
    bubbling, 31, 36–38, 41
    fractal, 38
Smooth cubic, 22–24
Software
    conversational multimedia system, 49
    3D-XplorMath, 1–3
    evolver, 163–164
    Pixar's RenderMan, 163
    Povray, 13
Soliton, 5

T
Thickening, 42
Topology, 153, 164
Torsion
    of a curve, 137, 138, 142
    null-torsion-surface, 138, 141, 143–151

V
Vector-field, 136, 140–144, 146–148, 150, 151

W
Weiestraß, 88, 89, 102–103
    $\wp$ functions, 102
    formula, 88, 89
Willmore energy, 156–160

Printing: Ten Brink, Meppel, The Netherlands
Binding: Stürtz, Würzburg, Germany